T0180101

Studies in Systems, Decision and Control

Volume 126

Series editor

Janusz Kacprzyk, Polish Academy of Sciences, Warsaw, Poland
e-mail: kacprzyk@ibspan.waw.pl

About this Series

The series "Studies in Systems, Decision and Control" (SSDC) covers both new developments and advances, as well as the state of the art, in the various areas of broadly perceived systems, decision making and control- quickly, up to date and with a high quality. The intent is to cover the theory, applications, and perspectives on the state of the art and future developments relevant to systems, decision making, control, complex processes and related areas, as embedded in the fields of engineering, computer science, physics, economics, social and life sciences, as well as the paradigms and methodologies behind them. The series contains monographs, textbooks, lecture notes and edited volumes in systems, decision making and control spanning the areas of Cyber-Physical Systems, Autonomous Systems, Sensor Networks, Control Systems, Energy Systems, Automotive Systems, Biological Systems, Vehicular Networking and Connected Vehicles, Aerospace Systems, Automation, Manufacturing, Smart Grids, Nonlinear Systems, Power Systems, Robotics, Social Systems, Economic Systems and other. Of particular value to both the contributors and the readership are the short publication timeframe and the world-wide distribution and exposure which enable both a wide and rapid dissemination of research output.

More information about this series at http://www.springer.com/series/13304

Vladimir D. Noghin

Reduction of the Pareto Set

An Axiomatic Approach

Vladimir D. Noghin
Department of Control Theory
Saint Petersburg State University
Saint Petersburg
Russia

ISSN 2198-4182 ISSN 2198-4190 (electronic)
Studies in Systems, Decision and Control
ISBN 978-3-319-88501-8 ISBN 978-3-319-67873-3 (eBook)
https://doi.org/10.1007/978-3-319-67873-3

© Springer International Publishing AG 2018
Softcover reprint of the hardcover 1st edition 2017
This work is subject to copyright. All rights are reserved by the Publisher, whether the whole or part of the material is concerned, specifically the rights of translation, reprinting, reuse of illustrations, recitation, broadcasting, reproduction on microfilms or in any other physical way, and transmission or information storage and retrieval, electronic adaptation, computer software, or by similar or dissimilar methodology now known or hereafter developed.
The use of general descriptive names, registered names, trademarks, service marks, etc. in this publication does not imply, even in the absence of a specific statement, that such names are exempt from the relevant protective laws and regulations and therefore free for general use.
The publisher, the authors and the editors are safe to assume that the advice and information in this book are believed to be true and accurate at the date of publication. Neither the publisher nor the authors or the editors give a warranty, express or implied, with respect to the material contained herein or for any errors or omissions that may have been made. The publisher remains neutral with regard to jurisdictional claims in published maps and institutional affiliations.

Printed on acid-free paper

This Springer imprint is published by Springer Nature
The registered company is Springer International Publishing AG
The registered company address is: Gewerbestrasse 11, 6330 Cham, Switzerland

Preface

Almost any kind of human activity is associated with the following situations: There exist several alternatives, and a human being is free to choose any most suitable for him.

Best choice problems form the subject of decision theory. Using the latter, it is possible to perform choice in a more reasonable way, taking advantage of available information about preferences. This theory allows eliminating wittingly unsuitable alternatives, with thorough consideration of possible negative consequences caused by half-baked choice.

An extremely wide and practically important class of choice problems concerns multicriteria choice problems, where the quality of an accepted alternative is assessed by several criteria simultaneously. Given a set of criteria, a common approach here employs the Edgeworth–Pareto principle stating that the "best" alternatives are the Pareto optimal ones only. However, the Pareto set is often large, and the final choice within it seems difficult. This aspect leads to the so-called Pareto set reduction problem. The solution to this problem, i.e., the well-justified reduction of the Pareto set, is impossible without information about the decision-maker's preferences. A major source of such information consists in the decision-maker's preference relation. In the elementary case, this information indicates which of two Pareto optimal alternatives is preferable for the decision-maker (DM). Such kind of information (known as an information quantum) serves for eliminating one of the two alternatives, thereby slightly facilitating further choice of the "final" alternative. By adopting some rather natural constraints (axioms) that regulate the DM's choice procedure, one can reduce much more Pareto optimal alternatives using a single quantum. At the same time, with a collection of information quanta available, one can expect to obtain a relatively narrow set appreciably simplifying the final choice.

Adhering to a rigorous form of exposition, the author still endeavored not to lose the connection between theory and practice, involving all available means for the informal discussion and visualization of all new notions and results.

This book is intended for the specialists in the field of decision-making, requiring a standard university course on mathematics from a reader. No doubt, this book will be useful for those who solve multicriteria problems by occupation, namely researchers, design engineers, product engineers, developers, analysts. In addition, this book can be of certain value for the undergraduates and postgraduates specialized in mathematics, economics, and engineering.

Section 5.4 was written by O.V. Baskov, while Sects. 5.2, 8.6, and Subsects. 7.4.3, 7.4.4 by A.O. Zakharov. Section 5.1 proceeded from the research work of O.N. Klimova, and Section 8.5 was written jointly with A.V. Prasolov. The author expresses his sincere gratitude to the listed colleagues for their contribution to this book.

The formulas, figures, and assertions have the double numbering system, where the first number corresponds to the chapter.

The symbols □ and ■ indicate the beginning and end of a proof, respectively.

The author is grateful to the Russian Foundation for Basic Research for supporting his investigations in the field since 1998. And finally, my deep appreciation is expressed to A. Yu. Mazurov for his careful translation, permanent feedback, and contribution to the English version of the manuscript.

Saint Petersburg, Russia Vladimir D. Noghin

Contents

Notations

$A \backslash B$	The set-theoretic difference of sets A and B
$A \times B$	The Cartesian product of sets A and B
R^m	The Euclidean space of real m-dimensional vectors
$0_m = (0, 0, \ldots, 0) \in R^m$	The origin (zero) of space R^m
R^m_+	The nonnegative orthant of space R^m (the set of all nonnegative vectors except the origin)
R_+	The set of positive real numbers
N^m	The set of all vectors from space R^m with at least one positive and at least one negative components
$\|A\|$	The number of elements in a finite set A
$\sup Z$	The least upper bound of a numerical set Z
$\inf Z$	The greatest lower bound of a numerical set Z
$[z]$	The integer part of a number z
$\langle a, b \rangle = \sum_{i=1}^{m} a_i b_i$	The scalar product of vectors $a = (a_1, a_2, \ldots, a_m)$ and $b = (b_1, b_2, \ldots, b_m)$
$\\|a\\| = \sqrt{\langle a, a \rangle}$	$= \sqrt{a_1^2 + a_2^2 + \ldots + a_m^2}$
a > b	$\Leftrightarrow a_i > b_i, i = 1, 2, \ldots, m$
$a \geqq b$	$\Leftrightarrow a_i \geqq b_i, i = 1, 2, \ldots, m$
$a \geq b$	$\Leftrightarrow a \geqq b$ and $a \neq b$
$\text{cone}\{a^1, a^2, \ldots, a^k\}$	The convex cone generated by vectors $a^1, a^2, \ldots, a^k$ (the set of all linear nonnegative combinations of these vectors)
m	The number of criteria
$I = \{1, 2, \ldots, m\}$	The index set of criteria
X	The set of feasible alternatives
$f = (f_1, f_2, \ldots, f_m)$	Vector criterion
$Y = f(X)$	The set of feasible vectors
$\succ_X$	The DM's preference relation defined on a set X
$\succ_Y$	The DM's preference relation induced by the relation $\succ_X$ and defined on a set Y

$\succ$	The extension of $\succ_Y$ to the whole space R^m
$C(X)$	The set of selectable alternatives
$C(Y)$	The set of selectable vectors
Ndom X	The set of nondominated alternatives
Ndom Y	The set of nondominated vectors
$P_f(X)$	The set of Pareto optimal (efficient) alternatives (domain of compromise)
$P(Y)$	The set of Pareto optimal (efficient) vectors

Introduction

Any optimization problem (extremum problem) contains two objects, namely a set of feasible alternatives X and a numerical function (criterion) f defined on this set. The solution to an extremum problem is an element of the set X attaining the largest (or least) value of the criterion. Therefore, an extremum problem represents a maximization problem or a minimization problem, and all results obtained for the problems of one type can be easily reformulated for the problems of the other types. To present, the theory of extremum problems has been developed intensively, yielding numerous solution methods and algorithms for certain classes of such problems.

Apparently, the multicriteria problems (i.e., the ones with several numerical criteria $f_1, f_2, \ldots, f_m$) first appeared in the implicit form in the seventeenth century as voting problems. Following the origination of mathematical economics in the nineteenth century, they were successfully applied to solve different economic problems. The notion of a Pareto optimal alternative (domain of compromise) was pioneered in that period, too. A Pareto optimal alternative possesses a remarkable property: It cannot be improved, viz. increased or decreased (depending on whether the original problem concerns maximization or minimization) at least in terms of one criterion without degrading the values of the rest of the criteria. This notion was introduced by F. Edgeworth in the 1850s in the case of two criteria and then generalized by V. Pareto at the junction of the nineteenth and twentieth centuries to the case of several criteria. The overwhelming majority of researchers believe that the "best" choice should be made from the Pareto optimal alternatives (the so-called Edgeworth–Pareto principle). However, note that this principle has been given a rigorous statement and justification just recently [37], formerly accepted as some obvious axiom.

Unfortunately, in most multicriteria problems the Pareto set (domain of compromise) is rather wide, making specific choice within it nontrivial. This circumstance leads to the Pareto set reduction problem associated with searching for a certain Pareto optimal alternative as the "best" one. The favorable solution to this problem forms a major practical interest, as in applications-relevant choice problems one should consider a single or a few alternatives.

Nowadays, there exists a wide range of approaches to solve the Pareto set reduction problem, from the heuristic to axiomatic ones. Nevertheless, none of them claims to be uniform and "perfect" (perhaps, such an approach would hardly be proposed). Each of the existing approaches combines unquestionable advantages with patent drawbacks.

The abstract choice problem is to indicate one or several alternatives (called selectable alternatives) from a given original set of feasible alternatives X. Denote by $C(X)$ the set of all selectable alternatives. This set has to be found as the result of problem solution. Clearly, $C(X) \subset X$; in particular, the set $C(X)$ possibly makes a singleton. Fix some set A containing X. A function associating each set $X \subset A$ with its subset $C(X)$ is called a choice function on A. Thereby, generally a choice function is defined on the one- and multielement subsets of A (including X). In a special case, the value of a choice function on some set can be an empty set ("refusal to choose"). Choice theory [1] studies the general properties of a choice function depending on the variation of the set X (and $C(X)$) within a certain fixed set A.

Meanwhile, in contrast to choice theory, we will analyze the problems with a fixed set X and a collection of numerical functions forming a vector criterion $f = (f_1, f_2, \ldots, f_m)$. By assumption, the choice is performed by some decision-maker (DM), either an individual or a group of people pursuing a definite goal. For the choice to fit goal attainment as much as possible (becoming "best" or "optimal" for the DM), it is necessary to consider the "tastes" and "preferences" of the DM reflecting somehow the whole sense of "best" alternative. In many cases, the DM "preferences" can be expressed using a binary (preference) relation $\succ_X$ defined on the set X. The notation $x' \succ_X x''$ reads as "the first alternative is preferable to the second one." This means that the DM chooses the first alternative and never the second from these two alternatives. As a matter of fact, by determining the preference relation $\succ_X$, one reduces the choice function to the two-element subsets of the set X. This relation is generally incomplete, and there may exist pairs of alternatives not interconnected by the relation $\succ_X$. In other words, the DM appears unable to give preference to any of the alternatives in such pairs.

A triplet $\langle X, f, \succ_X \rangle$ is called a multicriteria choice model. The classical multicriteria optimization model (e.g., see [9, 10, 18, 19, 26, 53, 58–60, 64, 65]) does not incorporate the preference relation $\succ_X$, whereas general choice theory does not involve the vector criterion. Note that the framework of the classical multicriteria model containing merely X and f fails to formulate the problem in a rigorous way. For instance, while announcing that it is desirable to maximize (or minimize) the criteria, the researchers actually imply the existence of some binary relations expressing this statement strictly. Therefore, the multicriteria model must include a binary relation, although many authors still ignore this. A possible explanation is that, unlike X and f, the preference relation guiding the DM often appears fragmentary.

According to the Edgeworth–Pareto principle, the set of selectable alternatives $C(X)$ lies within the Pareto set and any Pareto optimal alternative can be chosen. Using the set X and the vector criterion f, it is possible to construct the Pareto set, perhaps with serious computational difficulties. The next stage is to choose an alternative within the Pareto set or, in other words, to reduce this set to acceptable sizes. Consequently, under the Edgeworth–Pareto principle, the multicriteria choice problem can be formulated as the Pareto set reduction problem to the set of selectable alternatives $C(X)$. This problem seems so complicated that even called the Pareto set reduction challenge [30, 39].

The Pareto set can be reduced only with some additional information about the multicriteria problem (besides X and f). In many cases, investigators replace such information with certain heuristic considerations or definite "plausible" hypotheses that narrow the search space of "best" (selectable) alternatives. A distinctive feature of the heuristic methods is that there exists no clear description for the class of multicriteria choice problems, where these methods surely yield a desired result. In this context, the axiomatic approaches are more justified, since axioms define the class of associated problems, where a desired result is surely obtained.

A number of authors believe that the final choice must be performed by the DM after the direct analysis of the whole Pareto set or its considerable part. Really, given a few Pareto optimal alternatives (ideally, two), one of them can be in principle chosen by comparing and studying their advantages and shortcomings. However, even in the case of two alternatives, the DM may get into an awkward predicament, e.g., when the number of criteria is huge. For a rather large Pareto set (not to mention the infinite set), the direct analysis of the Pareto optimal alternatives becomes difficult, and the successful solution to the problem requires some formalized procedure.

This book has sequential exposition based on the formal definition and further usage of the notion of an information quantum about a preference relation. The existence of such a quantum means the DM's willingness to compromise by neglecting a group of criteria in order to gain in terms of another group of criteria. The above definition has a very simple logic, being clear to the experts and, moreover, to the individuals involved in the decision-making process without deep knowledge of higher mathematics. This aspect seems substantial, as such information often comes from the latter.

Using the definition of an information quantum and its elementary properties, it is possible to deal with the main question behind this notion: How should such information be taken into account? As it has turned out, the information quanta can be easily considered if one slightly restricts the class of the multicriteria choice problems that satisfy the Edgeworth–Pareto principle by adding several rather sensible requirements (axioms) on the DM's preference relation. In particular, one should just recalculate the original vector criterion by definite formulas, obtaining the Pareto set in terms of the new vector criterion. Generally, the new Pareto set is narrower than that of the original problem and, what is more important, all selectable alternatives of the original problem stay within the new Pareto set. In other words, the transition from the original Pareto set to the new one actually reduces the domain of compromise without losing the selectable alternatives. Owing to the consideration of the existing information about the preference relation, the search space of the selectable alternatives becomes narrower, thereby simplifying the choice problem.

The aforesaid is the content of the first three chapters. Chapter 4 introduces the notion of a consistent collection of information quanta and three assertions (criteria) to verify whether a given collection of information quanta enjoys consistency or not. Next, we find out how the so-called elementary collections of information quanta should be taken into account.

Chapter 5 analyzes the usage of different finite collections of information quanta and their application to Pareto set reduction. In particular, we describe two algorithms (the geometric and algebraic ones) for considering an arbitrary finite number of consistent information quanta.

Chapter 6 explores the completeness of a collection of information quanta. As demonstrated there, using only a finite collection of such quanta it is possible to construct an arbitrarily accurate (in some sense) approximation of the unknown set of nondominated alternatives in the form of the Pareto set for a certain new multicriteria problem. The results derived in Chap. 6 demonstrate the important role of the information about the preference relation in the form of quanta. This information is complete; i.e., for a sufficiently large class of the multicriteria choice problems with a finite set of feasible alternatives, such information suffices for producing a clear view of the set of nondominated alternatives.

Chapter 7 gives the definition of an information quantum in the case of a fuzzy preference relation (a fuzzy information quantum), formulating the corresponding axioms of “reasonable” fuzzy choice and establishing some results for taking into account different fuzzy information quanta.

The outcomes of the preceding chapters are then accumulated in Chap. 8 that describes explicitly the general axiomatic Pareto set reduction approach based on the information about the preference relation in the form of information quanta. This chapter begins with the psychological aspects of human decision-making. Next, the axiomatic method is formulated and discussed in detail. The underlying principle of the method can be elucidated via comparison with Michelangelo’s creative process. As is well-known, the great sculptor was asked how he managed to create masterpieces like David from shapeless stone. Michelangelo answered, “It is easy. You just chip away the stone that doesn’t look like David.” The axiomatic approach is based on the same idea: using the available information about the preference relation in the form of a collection of quanta, sequentially eliminate from the original set of feasible alternatives all Pareto optimal alternatives that cannot be selected according to this information. This process continues until one obtains a set of alternatives satisfying the DM.

A major advantage of the suggested axiomatic approach lies in the following. Using a collection of axioms, we outline the class of multicriteria choice problems for which none of the selectable alternatives is eliminated at each step of the Pareto set reduction process. Thereby, this collection of axioms clearly specifies the applicability limits of the approach.

Furthermore, note that the axiomatic approach can be used in combination with other well-known solution methods of the multicriteria problems. For instance, in Chap. 8 we discuss how the sequential reduction method of the domain of compromise is employed jointly with the weighted sum-based approach and goal programming ideas. The justification of the linear scalarization method (a rather widespread linear combination of criteria) and some other nonlinear ones for Pareto set reduction is given. Chapter 8 also applies the obtained results to some economic problems, describing several ways to generalize the axiomatic approach by weakening the original axioms.

Chapter 1
Edgeworth-Pareto Principle

This chapter introduces and discusses the basic notions of decision-making in a multicriteria environment, namely, the set of feasible alternatives, the vector criterion and the preference relation of a decision-maker. Here we formulate the multicriteria choice problem. In addition, Chap. 1 defines a pair of fundamentally important notions, the set of nondominated alternatives and the Pareto set, which are vital for the statement and rigorous substantiation of the Edgeworth-Pareto principle.

The statement and substantiation of the above principle form the central outcome of Chap. 1. As established below, the Edgeworth-Pareto principle should be applied to solve the multicriteria choice problems from a certain sufficiently large class. This class comprises the problems satisfying two definite requirements (axioms) that express the "reasonable" behavior of a decision-maker. An attempt to use the Edgeworth-Pareto principle beyond the class is risk-bearing, possibly yielding inadequate results.

1.1 Multicriteria Choice Problem

1.1.1 Set of Feasible Alternatives and Set of Selectable Alternatives

In everyday life, people always face situations when it is necessary to perform choice. For instance, we choose one or another good in a shop. To reach an intended destination in a city or country, we choose a route and mode of transport. A school leaver chooses a university for further education or a place for employment. High-level executives form the stuff of their companies or departments; choose one or another behavioral strategy; make business and economic decisions. In different fields of science and technology, specialists designing various devices

© Springer International Publishing AG 2018

V.D. Noghin, *Reduction of the Pareto Set*, Studies in Systems, Decision and Control 126, https://doi.org/10.1007/978-3-319-67873-3_1

and equipment, buildings, cars, aircrafts, etc. also seek to choose the best engineering or design decision. Bankers choose objects for investment, the economists of enterprises and companies elaborate an optimal economic development program, and so on.

This list of real-world problems can be continued. Let us confine ourselves to mentioning just these problems, in order to identify common elements inherent in any choice problem.

First of all, there must exist a given set of alternatives for further choice. Denote it by X and call the *set of feasible alternatives*. This set has minimum two elements to provide real choice. And there are no restrictions on the maximum number of feasible alternatives, which can be finite or infinite. The nature of these alternatives does not really matter: project decisions, behavioral patterns, political or economic strategies, development scenarios, short- or long-term plans, etc.

The choice proper is to indicate an alternative (called the selected alternative) among all feasible ones. Note that, in many cases, not a single alternative but a whole collection of alternatives is chosen, making some subset of the set of feasible alternatives X. An elementary example concerns the competitive selection of university entrants.

Designate by $C(X)$ the *set of selectable alternatives*. It represents the solution of the choice problem and can be any subset of the set X. Therefore, to solve the choice problem one has to find the set $C(X)$, $C(X) \subset X$. If the set of selectable alternatives contains no elements (i.e., empty), the choice proper disappears, as none of the alternatives is selected. Such a situation attracts little interest in terms of applications; and so, the above set must include at least one element. In some problems, it can be infinite.

A specific feature of the multicriteria choice problems is that, in contrast to standard (single-criterion) optimization, there exists no uniform definition of solution for all occasions (and it would hardly be suggested). This circumstance can be explained as follows: the solution of a multicriteria choice problem [the set $C(X)$] is constructed in the course of choice using different additional information (besides X and f), depending substantially on a decision-maker.

We call attention to the fact that the set of selectable alternatives has been given just a notation without a rigorous definition. Actually, such a definition is absent. As illustrated by further exposition, this fact in no way decreases the strictness of the approach. The only requirement applied to the above set is stated by Axiom 1: an alternative that is not selected from some pair must not be selected from the original set of feasible alternatives. And all crucial theorems involving the set of selectable alternatives state that an *arbitrary* set $C(X)$ satisfying Axiom 1 has inclusion in a specific-form Pareto set. In other words, these theorems give some upper estimate for a whole class of unknown sets of selectable alternatives within the limits of Axiom 1.

1.1.2 Decision-Maker

Choice process is impossible without an individual implementing it for personal goals. An individual performing choice and bearing full responsibility for the consequences is called a *decision-maker (DM)*.

As a rule, the DM's nature makes no matter for the solution of the choice problem. For instance, imagine a human as the DM; obviously, it represents a complex biological and social being. This being possesses a body of definite organization with different (perhaps, incompletely explored) biochemical, psychophysical, physiological and mental processes. However, it is unnecessary to consider the structural peculiarities of its skull or spine to choose, e.g., a certain economic strategy of a company. For choice process, the important factors include the experience of this human in economics, its future expectations of the company and company-related interests, etc. Consequently, speaking about the DM in the context of the choice problem, we mean not a human as a whole, but only its "part" and characteristics associated with choice process.

In decision theory, different individuals demonstrating the same behavior in same situations are indistinguishable from each other, representing the same DM.

1.1.3 Vector Criterion

According to standard considerations, the chosen (best) decision is a feasible alternative satisfying the wishes, interests or goals of a given DM as fully as possible. The DM's aspiration for a definite goal is often expressed in mathematical terms by maximization (or minimization) of some numerical function defined on the set X. But in difficult situations one deals with several such functions simultaneously. For example, imagine that a phenomenon, an object or a process is analyzed from different viewpoints each formalized using an appropriate numerical function. In other example for the dynamic (step-by-step) analysis of this phenomenon involving a dedicated function at each step, it is necessary to take into account several functions.

The book focuses on the case of several numerical functions $f_1, f_2, \ldots, f_m$, $m \geqq 2$, defined on the set of feasible alternatives X. Depending on the interpretation of the choice problem, these functions are called *optimality criteria*, *efficiency criteria*, *goal functions*, *quality indices*, or *performance criteria*. We emphasize that the criteria represent numerical functions, i.e., functions taking values in the real one-dimensional space.

Let us illustrate the terms by considering *the best project choice problem*. Here the set X consists of several competing projects (e.g., the construction of a new enterprise), while the optimality criteria can be the project costs f_1 and the expected profit f_2 from the implementation of a project (the profit yielded by an enterprise). With only one optimality criterion being treated in this problem, the practical importance of its solution is reduced to naught. Really, using the first criterion only, we choose the cheapest project, but its implementation possibly yields a very small profit. On the other hand, the implementation of the most profitable project (chosen by the second optimality criterion) often turns out impossible due to insufficient resources. This explains the need to use both optimality criteria simultaneously in the best project choice problem. Another goal can be associated with the minimization of the undesired environmental effects of an enterprise (its construction and further operation), which adds the third optimality criterion reflecting the ecological risks, and so on. And the role of the DM in this problem is played by the corresponding municipal administration (for a public-owned enterprise) or the executive officer of a parent company (for a private owned enterprise).

The numerical functions $f_1, f_2, \ldots, f_m$ form the *vector criterion*

$$f = (f_1, f_2, \ldots, f_m), \tag{1.1}$$

which takes values in the m-dimensional vector space R^m. This space is called the *criterion space*, and for some $x \in X$ the value $f(x) = (f_1(x), f_2(x), \ldots, f_m(x)) \in R^m$ of the criterion f is called a *feasible vector* associated with the alternative x. All feasible vectors make the *set of feasible vectors*

$$Y = f(X) = \{y \in R^m \mid y = f(x) \text{ for some } x \in X\}.$$

Along with the set of selectable alternatives, it seems convenient to introduce the *set of selectable vectors*

$$C(Y) = f(C(X)) = \{y \in Y \mid y = f(x) \text{ for some } x \in C(X)\},$$

representing a certain subset of the criterion space R^m.

1.1.4 Multicriteria Problem

A problem with a set of feasible alternatives X and a vector criterion f is usually called a *multicriteria* (*multi-objective* or *vector optimization*) *problem*. The properties of such problems were explored in a vast literature (e.g., see [9, 10, 18, 19, 26, 53, 58–60, 64, 65]).

Note that, in many cases, mathematical modeling of decision-making (i.e., designing the set X and the vector criterion f) represents a complicated process with close cooperation of two groups of specialists, namely, experts in a concrete field of knowledge associated with a real problem and experts in decision-making (mathematicians). On the one hand, it is necessary to consider all major features and details of the real problem. On the other hand, the resulting model shall be not very complex so that the up-to-date mathematical methods and procedures become applicable. Therefore, the modeling stage substantially depends on the experience, intuition and skills of both groups of specialists. It is impossible to identify this stage with the simple and formal usage of the well-known algorithms.

Another important aspect should be emphasized in this context as well. Any choice problem (including the multicriteria ones) has a close connection to a specific DM. Even at the stage of mathematical modeling that yields the set of feasible alternatives and the vector criterion, the specialists can't do without the advice, recommendations and instructions of the DM, as the vector criterion serves for expressing the DM's goals. Clearly, it is impossible to construct an ideal model reproducing all real circumstances. A model always gives a simplified description of the reality. What is important, the model must incorporate all features and details mostly affecting the final choice of the best alternative.

Assume that the two elements of the choice problem are formed, explicitly described and fixed. Evidence suggests that, in most cases, it is impossible to express the whole gamut of desires, tastes and preferences of a given DM in terms of the criterion f. In fact, the vector criterion merely indicates some local goals that often conflict with each other. As a rule, these goals cannot be achieved all at once, and additional information is necessary to implement a compromise. In other words, with the two elements only (the set of feasible alternatives and the vector criterion), the choice problem appears "underdetermined." This underdetermined character later results in the weak logical validity of the best choice.

The well-justified choice requires some additional information about the DM's preferences besides vector criterion. With this purpose, our idea is to introduce one more element in the multicriteria problem, which would express and describe these preferences.

1.1.5 Preference Relation

Consider two feasible alternatives, x' and x''. Assume that this pair of alternatives is presented to a DM, and the latter chooses (gives preference) to the first alternative. Then we write

$$x' \succ_X x''.$$

The sign $\succ_X$ denotes a *strict preference relation* or, in short, a *preference relation*.

Interestingly, it is possible that neither the relation $x' \succ_X x''$ nor the relation $x'' \succ_X x'$ holds for some feasible alternatives x' and x''. In other words, a DM is unable to make a final choice between them. Quite probably, there may exist pairs of different alternatives where a DM cannot give preference to one of them. The described situation agrees with the reality. Moreover, if a DM was required to choose a preferable alternative in an arbitrary pair of feasible alternatives, the theory based on this "stringent" requirement to the DM would be of no value in practice. Such "omnipotent" DMs are a very rare phenomenon in one's life.

A preference relation $\succ_X$ defined on the set of feasible alternatives naturally induces a preference relation $\succ_Y$ on the set of feasible vectors Y as follows[1]:

$$f(x') \succ_Y f(x'') \quad \Leftrightarrow \quad x' \succ_X x'', \quad \text{for}\, x', x'' \in X.$$

And so, the vector $y' = f(x')$ is preferable to the vector $y'' = f(x'')$ (i.e., $y' \succ_Y y''$) if and only if the alternative x' is preferable to the alternative x'' (i.e., $x' \succ_X x''$).

1.1.6 *Multicriteria Choice Problem*

It is now possible to formulate all basic elements of the multicriteria choice problem. Thus, the statement of any *multicriteria choice problem* includes

- the set of feasible alternatives X,
- a numerical vector criterion f of form (1.1),
- a preference relation $\succ_X$ defined on the set of feasible alternatives.

The multicriteria choice problem statement omits the DM proper: there is no need for this. By assumption, the vector criterion and the preference relation "materialize" the aspirations, tastes, desires and preferences of the DM affecting the process of its choice.

However, we emphasize that the basic elements of a multicriteria choice problem can be augmented if necessary by new objects deeper reflecting the interests, motivation and desires of the DM. But this situation is not studied in the book.

The above multicriteria choice problem is stated in terms of alternatives. In many cases, in terms of vectors it is more convinient. Then it contains two objects, namely,

- the set of feasible vectors Y, $Y \subset R^m$,
- a preference relation $\succ_Y$ defined on the set of feasible vectors.

1.2 Binary Relations

1.2.1 *Definition of Binary Relation*

There exists a special mathematical notion, called binary relation, for describing and studying the preference relation defined earlier. Section 1.2.1 presents an

[1]Some different alternatives may have same vector values. And so, it seems correct to replace "for x', $x'' \in X$" with "for all $x_1 \in \tilde{x}_1$, $x_2 \in \tilde{x}_2$, $\tilde{x}_1, \tilde{x}_2 \in \tilde{X}$, where $\tilde{X}$ represents the aggregate of the equivalence classes induced by the preference relation $x_1 \sim x_2 \Leftrightarrow f(x_1) = f(x_2)$ on the set X." Here $\tilde{x}_i$ denotes the equivalence class induced by the element x_i, $i = 1, 2$.

auxiliary mathematical framework connected with binary relations. A reader having a good knowledge of the subject may glance over it, passing to the next section.

First, recall the notion of the Cartesian product of two sets. Consider two nonempty arbitrary sets A and B. Their *Cartesian product* is a set $A \times B$ defined by

$$A \times B = \bigcup_{\substack{a \in A \\ b \in B}} \{(a, b)\}.$$

That is, the Cartesian product of two sets consists of all possible pairs of their elements such that the first and second elements in a pair are elements of the first and second sets, respectively.

For example, the Cartesian product of the two finite numerical sets $A = \{1,\ 2\}$ and $B = \{2,\ 3,\ 4\}$ contains six elements, having the form

$$A \times B = \{(1,\ 2), (1,\ 3), (1,\ 4), (2,\ 2), (2,\ 3), (2,\ 4)\}.$$

A *binary relation* $\Re$ defined on a set A is a subset of the Cartesian product $A \times A$, i.e., $\Re \subset A \times A$. In other words, any set of pairs composed of elements from a set A actually forms some binary relation. Particularly, the "largest" binary relation is the set $\Re = A \times A$ coinciding with the Cartesian product of A by itself.

If the inclusion $(a, b) \in \Re$ holds, in a standard fashion we write $a \Re b$ and say that *element a is in relation* $\Re$ *with element b*(equivalently, *element a dominates element b in terms of relation* $\Re$). When both inclusions $a \Re b$ and $b \Re a$ fail, we say that *these elements are incomparable* in terms of relation $\Re$.

Generally, $a \Re b$ does not imply $b \Re a$.

Consider the examples of some binary relations. The school curriculum covers a series of binary relations defined on the real space, viz., =, $\geqq$, $\leqq$, >, and <. Set theory deals with the inclusion relation $\subset$ defined on the set of all subsets of a fixed set.

For arbitrary vectors $a = (a_1, a_2, \ldots, a_m)$ and $b = (b_1, b_2, \ldots, b_m)$ from the criterion space R^m, let us introduce the following binary relations that will be involved throughout this book:

$$\begin{aligned} a > b &\Leftrightarrow \quad a_i > b_i, \quad i = 1, 2, \ldots, m, \\ a \geqq b &\Leftrightarrow \quad a_i \geqq b_i, \quad i = 1, 2, \ldots, m, \\ a \geq b &\Leftrightarrow \quad a \geqq b \text{ and } a \neq b. \end{aligned}$$

The last relationship $a \geq b$ means that each component of vector a is greater or equal to the corresponding component of vector b and at least one component of the first vector is strictly greater than the corresponding component of the second vector. The binary relation $\geq$ is often called the *Pareto relation.*

1.2.2 Types of Binary Relations

The typification of binary relations depends on their properties. In what follows, we give definitions for some widespread types of binary relations.

A binary relation $\Re$ defined on a set A is called

- *reflexive* if the relationship $a\,\Re\,a$ holds for all elements $a \in A$;
- *irreflexive* if the relationship $a\,\Re\,a$ fails for all elements $a \in A$;
- *symmetrical* if every time the relationship $a\,\Re\,b$ holding for some elements $a, b \in A$ implies the relationship $b\,\Re\,a$;
- *asymmetrical* if every time the relationship $a\,\Re\,b$ holding for some elements $a, b \in A$ implies that the relationship $b\,\Re\,a$ fails;
- *antisymmetric* if every time the relationships $a\,\Re\,b$ and $b\,\Re\,a$ holding for some elements $a, b \in A$ imply their equality $a = b$;
- *transitive* if, for any triplet of elements $a,\ b,\ c \in A$, the relationships $a\,\Re\,b$ and $b\,\Re\,c$ imply the relationship $a\,\Re\,c$;
- *invariant with respect to a linear positive transformation* if, for any triplet of elements $a,\ b,\ c \in A$ and an arbitrary positive number α, the relationship $a\,\Re\,b$ implies the relationship $(\alpha \cdot a + c)\ \Re\ (\alpha \cdot b + c)$ (here $A = R^m$);
- *complete* if for any pair of elements $a,\ b \in A,\ a \neq b$, we have the relationship $a\,\Re\,b$ or the relationship $b\,\Re\,a$;
- *partial* if this relation is not complete; then there exists a pair of elements in the set A that are incomparable in terms of the relation $\Re$.

The equality relation = and the nonstrict inequality relation $\geqq$ are reflexive, while the strict inequality relation > and the relation $\geq$ are irreflexive on R^m. Moreover, the equality relation and the nonstrict inequality relations are symmetrical and antisymmetric, respectively; the relations > and $\geq$ are asymmetrical. And all the relations mentioned, $=,\ \geq,\ >$, and $\geqq$, are transitive and invariant with respect to a linear positive transformation. Obviously, the equality relation on a set of numbers or vectors and the inclusion relation are partial. At the same time, the nonstrict inequality relation considered on the real vector space R^m with $m > 1$ becomes partial only, too.

As easily verified, *any asymmetric relation is irreflexive.*

□ Really, assume that a certain asymmetric relation $\Re$ is not irreflexive. Then for some $a \in A$ we have the relationship $a\Re a$. Due to the asymmetrical property of $\Re$, this relationship takes no place. The resulting contradiction leads to the irreflexivity of $\Re$. ■

1.2.3 Ordering Relations

The combinations of some binary relations play an important role in set theory. Let us introduce corresponding definitions.

A binary relation $\Re$ defined on a set A is called

- *a nonstrict order* (*a nonstrict ordering relation*) if it is reflexive, antisymmetric and transitive;
- *a strict order* (*a strict ordering relation*) if it is irreflexive and transitive;
- *a linear order* if it is a complete nonstrict or complete strict order.

The inequality relations $\geqq$ and $>$ represent a linear order on the real space, yet making no sense on the real vector space. The relation $\geq$ considered on the real vector space is a strict partial order.

Lemma 1.1 *Any strict ordering relation is asymmetrical* □

Again, we prove by contradiction: suppose that a certain relation $\Re$ is irreflexive and transitive, but not asymmetrical. And so, there exists a pair of elements $a,\ b \in A$ such that the relationships $a \Re b$ and $b \Re a$ hold simultaneously. By transitivity, this directly implies $a \Re a$, which is inconsistent with the irreflexivity of the relation $\Re$. ■

Another example of a strict linear order defined on space R^m is given by the following lexicographical ordering relation. A vector $y' = (y'_1,\ y'_2, \ldots,\ y'_m)$ is *lexicographically greater* than a vector $y'' = (y''_1,\ y''_2, \ldots,\ y''_m)$ if and only if one of the following conditions hold:

(1) $y'_1 > y''_1$;
(2) $y'_1 = y''_1,\ y'_2 > y''_2$;
(3) $y'_1 = y''_1,\ y'_2 = y''_2,\ y'_3 > y''_3$;
....
(m) $y'_i = y''_i,\ i = 1, 2, \ldots, m-1;\ y'_m > y''_m$.

Clearly, any two vectors from space R^m coincide or one of them is lexicographically greater than the other. That is to say, the lexicographical ordering relation appears complete. Moreover, it is a transitive relation.

1.3 Exclusion Axiom and Set of Nondominated Alternatives

1.3.1 Asymmetry Requirement for Preference Relation

Consider the multicriteria choice problem with a set of feasible alternatives X, a vector criterion f and a preference relation $\succ_X$. As this preference relation is defined on the pairs of feasible alternatives, it obviously represents some binary relation.

The original approach suggested below evolves from the preference relation $\succ_X$, which is per se a strict preference relation in the sense that the relationship $x \succ_X x'$ negates the inverse relationship $x' \succ_X x$. In terms of binary relations discussed in the previous section, this means that the preference relation must be asymmetrical.

Therefore, in the sequel we will study the choice problems with the preference relations satisfying asymmetry. P. Fishburn [11]–[12] believes that asymmetry is the minimal requirement to the preference relation, a must-have among other requirements or conditions imposed on the DM's preference relation.

1.3.2 Exclusion Axiom

Consider two arbitrary feasible alternatives, x' and x''. By asymmetry, one and only one of the following cases takes place:

- The relationship $x' \succ_X x''$ holds but the relationship $x'' \succ_X x'$ fails;
- The relationship $x'' \succ_X x'$ holds but the relationship $x' \succ_X x''$ fails;
- Both relationships $x' \succ_X x''$ and $x'' \succ_X x'$ fail.

In the first case, i.e., under $x' \succ_X x''$, we say that the alternative x' *dominates* the alternative x'' (in terms of the relation $\succ_X$). In the second case, x'' *dominates* x'. And in the third case, we say that the alternatives x' and x'' *are incomparable* in terms of the preference relation.

Now, get back to the choice problem. Assume that, for some alternative x'', there exists an alternative x' such that the relationship $x' \succ_X x''$ takes place. According to the definition of the preference relation, the DM chooses the first alternative from this pair. And then the second alternative x'' cannot be selected from the pair x', x'' (otherwise, we obtain the relationship $x'' \succ_X x'$ contradicting together with $x' \succ_X x''$ the asymmetry of the relation $\succ_X$). In terms of the set of selectable alternatives, the aforesaid can be expressed as the equivalence

$$x' \succ x'' \Leftrightarrow C(\{x',\ x''\}) = \{x'\},$$

for all $x',\ x'' \in X$.

Imagine that the second alternative x'' is not chosen from the pair, since the latter contains a better alternative. Considering x'' within the whole set of feasible alternatives X, it seems natural to expect that x'' would not be chosen from the set X, too (as there exists at least one alternative x' belonging to the set X that is preferable to x'').

Following this line of reasoning, the second alternative in the pair x', x'' is not chosen from the whole set X if the first alternative is chosen from this pair. Thereby, throughout the book we accept

Axiom 1 (*the exclusion axiom of dominated alternatives*) If the relationship $x' \succ_X x''$ holds for a certain pair of alternatives $x', x'' \in X$, then $x'' \notin C(X)$.[2]

Axiom 1 involves the preference relation $\succ_X$ guiding the DM and, in addition, the set $C(X)$. Hence, this requirement should be interpreted as a definite constraint

[2]As easily verified, *the inverse Condorcet condition* [1] implies Axiom 1, but not vice versa.

on the set of selectable alternatives we will deal with. That is, any set of selectable alternatives must not contain such elements having preferable alternatives.

Now, we give a simple illustrative example where the exclusion axiom fails. Consider the choice problem with three candidates for two vacancies. By assumption, both vacancies must be filled. Direct comparison of the candidates shows that the first is preferable to the second and to the third, while the second is preferable to the third. Since two candidates must be chosen anyway, they are the first and the second candidates. Thus, the second candidate is not chosen from the pair of the first and second ones, yet being chosen from the whole set of three candidates. And the exclusion axiom of dominated alternatives takes no place in this example.

1.3.3 Set of Nondominated Alternatives

According to Axiom 1, any dominated alternative does not belong to the set of selectable alternatives. The elimination of all dominated alternatives from X yields a set playing a crucial role for further exposition.

The set of nondominated alternatives is defined by

$$\mathrm{Ndom}\, X = \{x^* \in X |\, \text{there exists no}\, x \in X \,\text{such that}\, x \succ_X x^*\}.$$

Therefore, Ndom$X$ represents a certain subset of the set of feasible alternatives X. Depending on the structure of the set X and a specific preference relation $\succ_X$, the set of nondominated alternatives may

- be empty (containing no elements);
- consist of one element (being a singleton);
- contain a finite number of elements;
- consist of infinitely many elements.

Lemma 1.2 *For any set of selectable alternatives $C(X)$ satisfying Axiom* 1, *we have the inclusion*

$$C(X) \subset \mathrm{Ndom}X. \tag{1.2}$$

□ Inclusion (1.2) holds for the empty set C(X). Suppose that inclusion (1.2) fails for some nonempty set $C(X)$. Then there exists an element $x'' \in C(X)$ such that $x'' \notin \mathrm{Ndom}\, X$. By the definition of the set of nondominated alternatives, there exists an alternative $x' \in X$ satisfying the relationship $x' \succ_X x''$. Then Axiom 1 immediately yields $x'' \notin C(X)$, which contradicts the hypothesis that x'' is the selectable alternative. ■

Inclusion (1.2) establishes that **choice should be performed only within the nondominated alternatives** for a sufficiently large class of the problems (the ones satisfying Axiom 1), and also any nondominated alternative can be selected.

If $C(X) \neq \varnothing$ and the set of nondominated alternatives represents a singleton, the choice problem is solved in principle: the unique element of this set must be chosen due to inclusion (1.2). However, note that such situations are very rare in practice. Nevertheless, even incomplete (fragmentary) information about the DM's preference relation allows to eliminate the dominated alternatives (unfit for choice) from the whole set of feasible alternatives, making further choice simpler.

Along with the set of nondominated alternatives, let us introduce *the set of nondominated vectors*

$$\begin{aligned}\operatorname{Ndom} Y &= f(\operatorname{Ndom} X) = \{f(x^*) \in Y \mid \text{there exists no}\, x \in X \,\text{such that}\, x \succ_X x^*\} \\ &= \{y^* \in Y \mid \text{there exists no}\, y \in Y \,\text{such that}\, y \succ_Y y^*\}.\end{aligned}$$

Axiom 1 and Lemma 1.2 can be reformulated for the set of nondominated vectors as follows.

Axiom 1 (*the exclusion axiom of dominated vectors*) If the relationship $y' \succ_Y y''$ holds for a certain pair of vectors $y',\ y'' \in Y$, then $y'' \notin C(Y)$.

Lemma 1.2 (in terms of vectors) *For any set of selectable vectors $C(Y)$ satisfying Axiom* 1, *we have the inclusion*

$$C(Y) \subset \operatorname{Ndom} Y.$$

1.4 Edgeworth-Pareto Principle

1.4.1 Pareto Axiom

The DM's interest in obtaining the maximum possible values for all components of the vector criterion f can be expressed using the Pareto axiom.

Pareto axiom (in terms of alternatives). *For all pairs of alternatives $x',\ x'' \in X$ satisfying the inequality $f(x') \geq f(x'')$, we have the relationship $x' \succ_X x''$.*

Recall that $f(x') \geq f(x'')$ means the component-wise inequalities $f_i(x') \geqq f_i(x'')$ for all $i = 1,\ 2, \ldots,\ m$, and also $f(x') \neq f(x'')$.

1.4.2 Pareto Set and Pareto Principle

If the inequality $f(x') \geq f(x'')$ holds for a certain pair of feasible vectors, then by the Pareto axiom the first alternative is preferable to the second one, i.e., $x' \succ_X x''$. According to Axiom 1, the second alternative would be chosen under no circumstances, and it can be eliminated from subsequent analysis in the decision-making process. The elimination of all such alternatives yields the Pareto set.

The set of Pareto optimal alternatives (domain of compromise) $P_f(X)$ is defined by

$$P_f(X) = \{x^* \in X| \text{ there exists no } x \in X \text{ such that } f(x) \geq f(x^*)\}.$$

Similarly, *the set of Pareto optimal vectors* $P(Y)$ is defined by

$$P(Y) = f(P_f(X)) = \{y^* \in Y \mid \text{ there exists no } y \in Y \text{ such that } y \geq y^*\},$$

where Y (as before) denotes the set of feasible vectors, i.e., $Y = f(X)$.

Edgeworth-Pareto principle *Under the exclusion axiom and the Pareto axiom, for any set of selectable alternatives* $C(X)$ *we have the inclusion*

$$C(X) \subset P_f(X). \tag{1.3}$$

□ Assume the opposite, i.e., for the alternative $x \in C(X)$ we have $x \notin P_f(X)$. Then, by the definition of the Pareto optimal alternative, there exists $x' \in X$ such that $f(x') \geq f(x)$. According to the Pareto axiom, this directly implies $x' \succ_X x$, and the exclusion axiom applied to this relationship yields $x \notin C(X)$, which contradicts the initial assumption $x \in C(X)$. ■

Inclusion (1.3) states *the Edgeworth-Pareto principle (the Pareto principle)*, namely, **if the DM demonstrates rather "reasonable" behavior (i.e., obeying the exclusion axiom and the Pareto axiom), then the alternatives chosen by the DM must be Pareto optimal.** Moreover, **any Pareto optimal alternative may be chosen under certain circumstances**.

This principle elucidates the special, extremely important role of the Pareto set in multicriteria decision-making. In particular, inclusion (1.3) indicates that the set of selectable alternatives $C(X)$ is the result of Pareto set reduction. Consequently, there are no selectable alternatives beyond the Pareto set; hence, search for the "best" alternatives can be immediately restricted to the limits of the Pareto set. This fact leads to *the Pareto set reduction problem*, i.e., the problem to find the selectable alternatives within the Pareto set.

Note that the Edgeworth-Pareto principle holds for a very wide class of multicriteria choice problems satisfying the two axioms above. For instance, it does not require the transitivity of the preference relation, which is a common property in many research works on decision theory.

1.4.3 Minimality of Exclusion Axiom and Pareto Axiom

An attempt to omit at least one of these two axioms can make the Edgeworth-Pareto principle invalid. As shown by the examples below, the exclusion axiom and the

Pareto axiom form the minimal collection of requirements guaranteeing the force of this principle.

Example 1.1 Consider $Y = \{y^1, y^2\}$, where $y^1 = (a, a)$, $y^2 = (b, a)$, $a > b$, and $y^2 \succ_Y y^1$. Let $C(Y) = \{y^2\}$. The exclusion axiom holds and $P(Y) = \{y^1\}$. But the Edgeworth-Pareto principle $C(Y) \subset P(Y)$ fails, since the Pareto axiom takes no place.

Example 1.2 Consider $Y = \{y^1, y^2\}$, where $y^1 = (a, a)$, $y^2 = (b, a)$, $a > b$, and $y^1 \succ_Y y^2$. Here $P(Y) = \{y^1\}$. Let $C(Y) = \{y^2\}$. The Pareto axiom holds, but the inclusion $C(Y) \subset P(Y)$ fails due to the violation of the exclusion axiom.

Most investigators do not prove inclusion (1.3). As a matter of fact, they believe that this inclusion is the axiom to-be-accepted. Such an approach involves not two but only one axiom. However, this "axiom" (in the form of the Edgeworth-Pareto principle) is not an axiom in the common sense, i.e., "an intuitively clear assertion taken on faith." It appears too "complicated." At the same time, the exclusion axiom and the Pareto axiom formulated in terms of the pairs of alternatives and vectors are substantially simpler for comprehension for everybody (including the DM). Hence, it is easier to test them in practice.

Moreover, the involvement of the two axioms (instead of one) seems useful in the following case. Some choice procedures proceed from the assumption (their authors believe in) that it is not necessary to choose from the Pareto set. In this context, the final choice of an alternative that is not Pareto optimal would surely violate one of them (the exclusion axiom or the Pareto axiom) or even both. The described circumstance must be taken into account by those who make the choice beyond the Pareto set.

1.5 Axioms of Transitivity and Compatibility

1.5.1 Transitivity Axiom

Consider situation where one alternative is preferable to another and the latter is in turn preferable to a third alternative. In such circumstances, a sensible individual comparing the first and third alternatives surely chooses the former. This situation resembles the congruence of numbers using the strict inequality relation. For instance, if $5 > 3$ and $3 > 1$, we surely have $5 > 1$. In terms of feasible alternatives, this property can be reformulated as follows: for any triplet x', x'', x''' of feasible alternatives, the relationships $x' \succ_X x''$ and $x'' \succ_X x'''$ surely imply the relationship $x' \succ_X x'''$. With regard to binary relations, this means that the preference relation adopted in the multicriteria choice problems must obey transitivity.

As emphasized earlier, we suppose that the preference relation is asymmetrical. According to the above considerations, introduce the condition (requirement) applied to the binary preference relations studied in this book: *the preference relation* $\succ_X$ *guiding the DM's choice is asymmetrical and transitive*. Note that an irreflexive and transitive relation is asymmetrical; and so, asymmetry can be replaced with the weaker requirements of transitivity and irreflexivity.

Recall that the preference relations $\succ_X$ and $\succ_Y$ are interconnected: the relationship $x \succ_X x'$ holds for feasible alternatives if and only if the relationship $f(x) \succ_Y f(x')$ holds for the corresponding vectors.

Next, we slightly "extend" the DM's hypothetical capabilities to compare vectors with each other. That is, assume that the DM can in principle compare any two vectors of the criterion space, not just the elements of the set Y.

And so, in the sequel we accept the following assumption formulated in terms of vectors from the criterion space.

Axiom 2 (transitivity of preference relation)[3] *The relation* $\succ_Y$ *(ergo, the relation* $\succ_X$*) is transitive. In addition, there exists the extension* $\succ$ *of the relation* $\succ_Y$ *to the whole criterion space* R^m *that is also transitive.*

By this axiom, any two vectors y', $y'' \in R^m$ satisfy one and only one of the following relationships:

- $y' \succ y''$;
- $y'' \succ y'$;
- neither $y' \succ y''$, nor $y'' \succ y'$ take place.

And the preference relation $\succ$ on the set of feasible vectors Y coincides with the relation $\succ_Y$ (in this case, the relation $\succ_Y$ is the reduction of $\succ$ to the set Y).

It is necessary to emphasize that there may exist a whole set of the extensions mentioned in Axiom 2, not just a unique one. As follows from a thorough study, the results obtained in the book are invariant to the choice of such extensions.

1.5.2 Compatibility Axiom

The multicriteria choice problem statement employs the vector criterion $f = (f_1, f_2, \ldots, f_m)$. As a rule, each component f_i of the vector criterion characterizes a definite goal of the DM, and the drive for this goal is often expressed in mathematical terms by maximization (or minimization) of the function f_i on the set X.

Note that several problems involve criteria that are not maximized or minimized. For instance, sometimes it is required to average the criterion in a certain sense or "keep" its value within definite limits, etc. In such situations, a more flexible approach is to replace the criteria f_i with the so-called "partial" preference relations $\succ_i$ (see [11–12] for details). However, as established in [12], in many applications-relevant cases (i.e., under certain "reasonable" requirements to $\succ_i$ and

[3]Note, that for the Edgeworth-Pareto principle to be true (see Theorem 1.1), in this axiom we may consider the extension $\succ$ not on the whole space R^m but merely on the Cartesian product $Y_1 \times Y_2 \times \ldots \times Y_m$, where Y_i is the smallest interval including $f_i(x)$, $i = 1, 2, \ldots, m$.

X) there exists a utility function u_i yielding an adequate description for this "individual" preference relation. In other words, for all $x', x'' \in X$ we have the equivalence $x' \succ_i x'' \Leftrightarrow u_i(x') > u_i(x'')$. These results show that many problems without criteria maximization (or minimization) in the original statement are, at least theoretically, reducible to such extremum problems.[4]

Obviously, in the multicriteria choice problem, the preference relation together with the optimality criteria expresses the interests of the same DM. Hence, they must match each other, be interconnected. The time is right to discuss this interconnection.

We say that *criterion f_i is compatible with the preference relation* $\succ$ if for any two vectors $y', y'' \in R^m$ such that

$$y' = (y'_1, \ldots, y'_{i-1}, y'_i, y'_{i+1}, \ldots, y'_m),\ y'' = (y'_1, \ldots, y'_{i-1}, y''_i, y'_{i+1}, \ldots, y'_m),\ y'_i > y''_i,$$

the relationship $y' \succ y''$ holds.

Conceptually, the compatibility of this criterion with the preference relation means that the DM is interested in the largest possible values of this criterion, other things being equal.

Let us express the interconnection between the preference relation of this DM and the optimality criteria in the form of the following requirement.

Axiom 3 (*criteria compatibility with preference relation*) Each of the criteria $f_1, f_2, \ldots, f_m$ is compatible with the preference relation $\succ$.

The Pareto axiom clearly implies the compatibility axiom, but the converse fails. If Axiom 3 is supplemented by the transitivity axiom, then the Pareto axiom can be guaranteed, as illustrated by the next result.

Lemma 1.3 *Axioms* 2 *and* 3 *imply the Pareto axiom.* □ Let the inequality $f(x') \geq f(x'')$ hold for two arbitrary feasible alternatives $x', x'' \in X$. Without loss of generality, assume that the strict inequality $f_k(x') > f_k(x'')$ take place for all indexes $k = 1, \ldots, l$ and some $l \in \{1, 2, \ldots, m\}$. For all subsequent indexes k, $k > l$ (if any, i.e., under the condition $l < m$), we believe that the corresponding equalities are satisfied.

By the compatibility of the first l criteria and the strict inequalities above, write

$$(f_1(x'), f_2(x'), \ldots, f_l(x'), \ldots, f_m(x')) \succ (f_1(x''), f_2(x'), \ldots, f_l(x'), \ldots, f_m(x')),$$
$$(f_1(x''), f_2(x'), \ldots, f_l(x'), \ldots, f_m(x')) \succ (f_1(x''), f_2(x''), f_3(x'), \ldots, f_l(x'), \ldots, f_m(x')),$$
$$\cdots\cdots\cdots\cdots\cdots\cdots\cdots\cdots$$
$$(f_1(x''), f_2(x''), \ldots, f_{l-1}(x''), f_l(x'), \ldots, f_m(x')) \succ (f_1(x''), f_2(x''), \ldots, f_l(x''), f_{l+1}(x'), \ldots, f_m(x')).$$

Hence, due to the transitivity of the preference relation $\succ$,

[4]We remark that further exposition of this chapter can be generalized to the case of "individual" preference relations.

$$(f_1(x'), f_2(x'), \ldots, f_l(x') \ldots, f_m(x')) \succ (f_1(x''), f_2(x'') \ldots, f_l(x''), f_{l+1}(x') \ldots, f_m(x')). \tag{1.4}$$

According to the earlier assumption, $f_k(x') = f_k(x'')$, $k = l+1, \ldots, m$. Therefore, relationship *(1.4) yields*

$$\begin{aligned} f(x') &= (f_1(x'), f_2(x'), \ldots, f_l(x'), \ldots, f_m(x')) \\ &\succ (f_1(x''), f_2(x''), \ldots, f_l(x''), \ldots, f_m(x'')) = f(x''), \end{aligned}$$

and, by the definition of the relation $\succ$, we arrive at the desired relationship $x' \succ_X x''$. ■

As established before, the acceptance of the exclusion axiom and the Pareto axiom guarantees the Edgeworth-Pareto principle. Based on the last lemma, it is possible to formulate a modification of this principle with three axioms instead of two. Taking into account that the image of a subset represents a subset of the image of the original set, we have the following result.

Theorem 1.1 (the Edgeworth-Pareto principle) *Within the conditions of Axioms* 1–3, *inclusions* (1.3) *and*

$$C(Y) \subset P(Y)$$

hold for any sets of selectable alternatives $C(X)$ *and selectable vectors* $C(Y)$, *respectively*.

Here is another fruitful result.

Lemma 1.4 *Under Axioms 2 and 3, the set of nondominated alternatives* Ndom X *satisfies the inclusion*

$$\text{Ndom}\, X \subset P_f(X). \tag{1.5}$$

□ On the contrary, let the relationship $x \notin P_f(X)$ be true for some nondominated element $x \in \text{Ndom}\, X$. Then, by the definition of the set of Pareto optimal alternatives, there exists an alternative $x' \in X$ such that $f(x') \geq f(x)$. By virtue of Lemma 1.3 and the premises of the current lemma, the Pareto axiom holds. Therefore, the resulting inequality implies the relationship $x' \succ_X x$, which is inconsistent with the initial hypothesis $x \in \text{Ndom}X$. ■

The subsets of the set of feasible alternatives (see above) under Axioms 1–3 are interconnected in the form of inclusions

$$C(X) \subset \text{Ndom}\, X \subset P_f(X) \subset X. \tag{1.6}$$

The widest set in relationship (1.6) is the set of feasible alternatives, whereas the narrowest one is the set of selectable alternatives. Figure 1.1 illustrates this interconnection graphically.

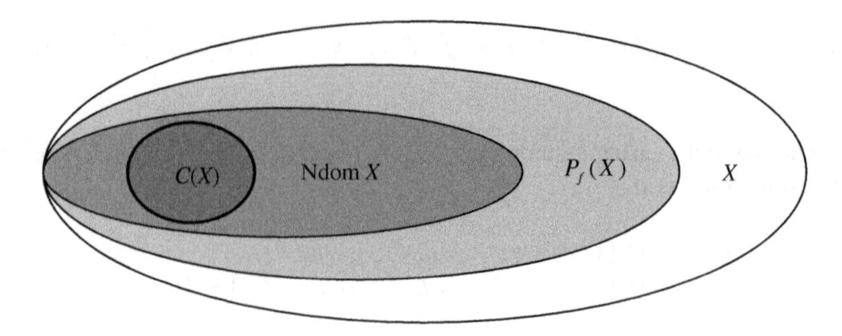

Fig. 1.1 Nested sets

In terms of vectors, inclusions (1.6) acquire the form

$$C(Y) \subset \operatorname{Ndom} Y \subset P(Y) \subset Y. \tag{1.7}$$

1.6 Finding of Pareto Set

1.6.1 Sets of Pareto Optimal Alternatives and Vectors

The equality $P(Y) = f(P_f(X))$ relates the sets of Pareto optimal alternatives and Pareto optimal vectors. Using this equality, one can find the set of Pareto optimal vectors based for a known set of Pareto optimal alternatives. The converse statement is true in some sense, too. In particular, given the set of Pareto optimal vectors $P(Y)$, a possible approach is to design the corresponding set of Pareto optimal alternatives by the formula $P_f(X) = f^{-1}(P(Y))$, where the right-hand side represents the preimage of the set $P(Y)$. Therefore, ideologically these two sets completely define each other, although an attempt to construct one of them from the other may encounter certain computational difficulties (in the first place, it applies to the Pareto optimal alternatives design).

Note that, in contrast to the arbitrary nature of the elements of the set $P_f(X)$, the elements of the Pareto set $P(Y)$ represent standard mathematical objects, viz. numerical vectors of dimension coinciding with the number of criteria m. And so, the set of Pareto optimal vectors seems more convenient for further consideration.

1.6.2 Calculation of Pareto Set

As is well-known, generally the Pareto set has a rather complex structure, often causing insurmountable difficulties in the course of its design and calculation.

In many applications, the set of feasible vectors Y (and the set of feasible alternatives X) contains a finite number of elements. In this case, the Pareto set $P(Y)$ can be constructed via the pairwise comparison of the elements belonging to the set Y in terms of the relation $\geq$, with the subsequent elimination of all dominated ones. This procedure yields

The Pareto set $P(Y)$ composed of at least one element.

□ Let us verify the last statement. Really, if each element of the set Y is dominated by a certain element from $Y_1 \subset Y$ in terms of the relation $\geq$, then by-turn each element of Y_1 must be dominated by a certain element from $Y_2 \subset Y_1$. And so on. This line of reasoning can be continued without any restrictions, and hence there exists an infinite sequence of elements $\{y^k\}_{k=1}^{\infty} \subset Y$ where a next element dominates the previous one. Due to the finite character of the set Y, this sequence includes identical elements. For instance, assume that $y^1 = y^k$ for some k. The equality $k = 1$ is ruled out by the asymmetry of the Pareto relation $\geq$. The equality $k = 2$ is impossible, as the second vector dominates the first (they do not coincide). If $k > 2$, then the transitivity of the Pareto relation gives the inequality $y^k \geq y^1$, which is inconsistent with the original equality $y^1 = y^k$. ■

As a matter of fact, we have established the following result.

Theorem 1.2 *In the case of a nonempty finite set of feasible vectors Y (in particular, if the set of feasible alternatives X is finite), there exists at least one Pareto optimal alternative and, accordingly, at least one Pareto optimal vector, i.e.,* $P_f(X) \neq \varnothing$, $P(Y) \neq \varnothing$.

Now, illustrate the Pareto set design procedure for the problem with four criteria.

Example 1.3 Let $m = 4$ and $Y = \{y^1, y^2, \ldots, y^5\}$. The feasible vectors are combined in the rows of Table 1.1.

First, to find the set of Pareto optimal vectors, let $Y_1 = Y$ and compare the first vector with the others. Obviously, all pairs

$$y^1, y^2; \quad y^1, y^3; \quad y^1, y^4; \quad y^1, y^5$$

appear incomparable in terms of the relation $\geq$. Therefore, memorize the vector y^1 as a Pareto optimal one and then eliminate it from the set Y_1.

Table 1.1 The feasible vectors

y^1	4	0	3	2
y^2	5	0	2	2
y^3	2	1	1	3
y^4	5	0	1	2
y^5	3	1	2	3

The resulting set is $Y_2 = \{y^2, y^3, y^4, y^5\}$. At step 2, compare the vector y^2 with the other elements of the set Y_2. The pair y^2, y^3 is incomparable in terms of the relation $\geq$. Since $y^2 \geq y^4$, eliminate the vector y^4 from the set Y_2. The remaining pair of vectors y^2, y^5 is incomparable in terms of the relation $\geq$. As the vector y^2 turns out nondominated, memorize it as a Pareto optimal one and then eliminate from the set Y_2.

The resulting set is $Y_3 = \{y^3, y^5\}$. Since $y^5 \geq y^3$, eliminate the vector y^3 to obtain only the vector y^5, which is also Pareto optimal.

The procedure has yielded the following set of Pareto optimal vectors: $P(Y) = \{y^1, y^2, y^5\}$.

1.6.3 Design Algorithm for Pareto Set

The design algorithm for the Pareto set can be rewritten in a better form for further programming. Consider the set of feasible vectors

$$Y = \{y^1, y^2, \ldots, y^N\}$$

composed of N elements.

Design algorithm for Pareto set $P(Y)$ includes seven steps as follows.

Step 1 Let, $i = 1$, $j = 2$, $P(Y) = Y$ forming the so-called *current set of Pareto optimal vectors*. At the beginning of the algorithm, this set coincides with the set Y, yielding the desired set of Pareto optimal vectors at the end. The algorithm is organized so that the desired set of Pareto optimal vectors comes out of Y through the sequential elimination of the surely nonoptimal vectors.

Step 2 Verify the inequality $y^i \geq y^j$. If true, proceed to Step 3; otherwise, move to Step 5.

Step 3 Eliminate the vector y^j from the current Pareto set $P(Y)$, as it is not Pareto optimal. Next, proceed to Step 4.

Step 4 Verify the inequality $j < N$. If true, let $j = j + 1$ and get back to Step 2; otherwise, move to Step 7.

Step 5 Verify the inequality $y^j \geq y^i$. If true, proceed to Step 6; otherwise, get back to Step 4.

Step 6 Eliminate the vector y^i from the current Pareto set $P(Y)$ and proceed to Step 7.

Step 7 Verify the inequality $i < N - 1$. If true, let sequentially $i = i + 1$ and $j = i + 1$. Then get back to Step 2. Otherwise (i.e., if $i \geqq N - 1$), finish the computations. The set of Pareto optimal vectors is completely constructed.

1.6.4 Geometry of 2D Pareto Set

Consider the elementary case with two criteria, $m = 2$. Here the set Y represents a certain ensemble of points on plane.

All points y satisfying the inequality $y \geq y^*$ form the angle with the vertex y^* and the arms parallel to the coordinate axes. But the point y^* does not belong to this angle, as $y \neq y^*$ (see Fig. 1.2).

Consider an example where the set of feasible points Y is a closed bounded domain (see Fig. 1.3).

To construct the set of Pareto optimal points $P(Y)$, take advantage of a geometrical observation adopted from Fig. 1.2. By the definition of the Pareto optimal vector y^*, there exist no points y such that $y \geq y^*$. Geometrically, all these points y form the angle with the vertex y^*. Hence, the point $y^* \in Y$ is Pareto optimal if and only if the corresponding angle with the vertex y^* and the arms parallel to the

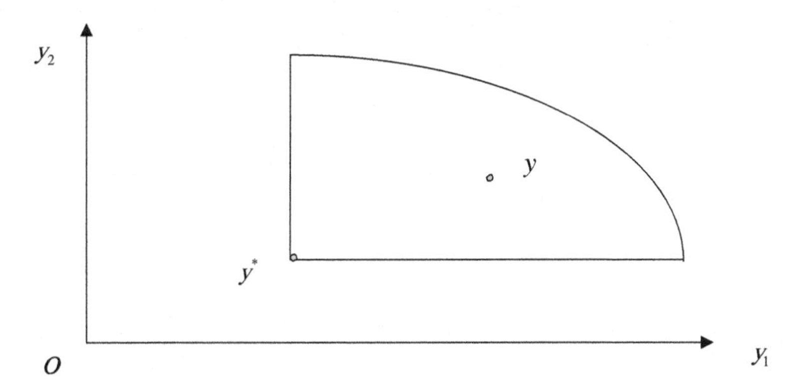

Fig. 1.2 Set of points y such that $y \geq y^*$

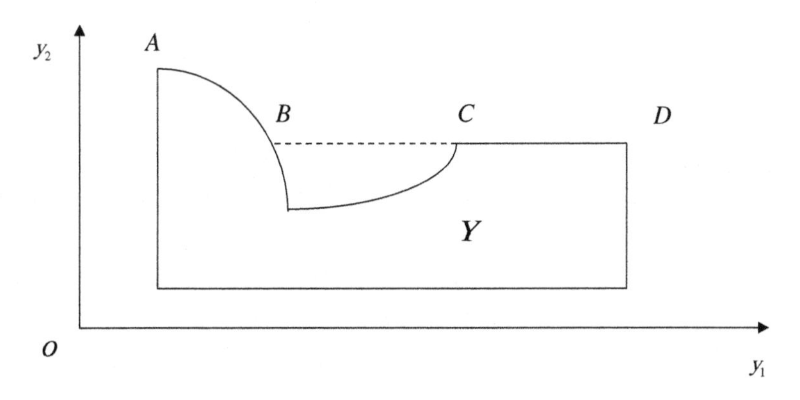

Fig. 1.3 Set Y

coordinate axes contains no points from the set Y. This means that the inner points of the set Y cannot be Pareto optimal. Among the boundary points of the set Y, the candidates for the Pareto optimal points are the ones located in the "northeastern" part (see the curve $ABCD$). Next, the boundary in the "dip" (the arc BC) does not belong to the Pareto set. And finally, among the segments of the northeastern boundary that are parallel to the coordinate axes, only the extreme point may be Pareto optimal; it is point D on the segment CD. As a result, we have obtained the following set of Pareto optimal points: the arch AB (except point B) and separate point D.

This geometrical approach to Pareto set design fails in the case of three or more criteria. Nevertheless, the modern computer-aided *visualization methods* [24] describe graphically the sets of feasible vectors and Pareto optimal vectors under relatively small m.

Chapter 2
Pareto Set Reduction Based on Elementary Information Quantum

The current chapter lays the foundation for the original axiomatic approach. First, we introduce the last (fourth) axiom on the invariance of preference relation. It is established that, within the accepted axiomatics, the DM's preference relation represents a cone relation with an acute convex cone. This feature allows employing the rich arsenal of convex analysis methods.

Next, we give the definition of an elementary information quantum about the unknown preference relation of the DM. The central result of the chapter is Theorem 2.5 that shows how the Pareto set can be reduced using an elementary information quantum.

In addition, different types of scales are discussed and the applicability of Theorem 2.5 to the multicriteria choice problems with criteria measured in arbitrary quantitative scales is justified.

2.1 Invariance Requirement for Preference Relation

2.1.1 Relations Invariant with Respect to Linear Positive Transformation

Recall the definition of an invariant relation given in Sect. 1.2. A binary relation $\Re$ defined on space R^m is called *invariant with respect to a linear positive transformation* if, for arbitrary vectors $y', y'' \in R^m$, any vector $c \in R^m$ and each positive number α, the relationship $y' \Re y''$ implies the relationship $(\alpha y' + c) \Re (\alpha y'' + c)$.

The inequality relations $>$, $\geqq$, $\geq$ defined on space R^m are the elementary examples of invariant binary relations. Obviously, a lexicographical relation (see Sect. 1.2) also belongs to the class of invariant binary relations.

In many application-oriented multicriteria choice problems, the preference relation $\succ$ can be considered invariant with respect to a linear positive

© Springer International Publishing AG 2018

V.D. Noghin, *Reduction of the Pareto Set*, Studies in Systems, Decision and Control 126, https://doi.org/10.1007/978-3-319-67873-3_2

transformation. Accordingly, let us supplement Axioms 1–3 by another one required for the development of a constructive mathematical theory.

Axiom 4 (preference relation invariance). *The preference relation $\succ$ is invariant with respect to a linear positive transformation.*

The invariance attributes of the relation $\succ$ are the properties of additivity and homogeneity. In other words, for any pair of vectors $y', y'' \in R^m$ such that $y' \succ y''$, the relationships $(y' + c) \succ (y'' + c)$ and $\alpha y' \succ \alpha y''$ hold for any vector $c \in R^m$ and any positive number α, respectively.

Lemma 2.1 *Owing to the transitivity and invariance of the preference relation $\succ$, the relationships $y \succ y'$ and $z \succ z'$ can be added termwise, i.e.,*

$$y \succ y', z \succ z' \Rightarrow y + z \succ y' + z'.$$

□ Add the vector z to both sides of the relationship $y \succ y'$. Using the additive property of the relation $\succ$, we obtain $y + z \succ y' + z$. The relationship $z \succ z'$ similarly implies $z + y' \succ z' + y'$. Now, taking advantage of the transitivity of the relation $\succ$, the relationships $y + z \succ y' + z$ and $z + y' \succ z' + y'$, we establish the desired result $y + z \succ y' + z'$. ■

2.1.2 *Cone Relations*

For further exposition, an important example of invariant binary relations is the class of cone relations. However, prior to the definition of a cone relation, we have to introduce some auxiliary notions of convex analysis.

A set A, $A \subset R^m$, is called *convex* if, together with any pair of points, it also contains the segment connecting them. In other words, a subset A of space R^m is convex if, for all pairs of points $y', y'' \in A$ and any number $\lambda \in [0, 1]$, we have the relationship $\lambda y' + (1 - \lambda) y'' \in A$. A set $K, K \subset R^m$, is called a *cone* if the inclusion $\alpha \cdot y \in K$ holds for each point $y \in K$ and any positive number α. A cone that is convex is called a *convex cone*. In other words, a convex set is a *convex cone* if, together with each point, it also contains the whole ray coming from the origin (generally, except the origin) to the given point. The origin (the *vertex* of a cone) may belong to this cone or not. As easily verified, the sum of any two (and more) elements of a convex cone belongs to this cone. A cone K is called *acute* if there exists no nonzero vector $y \in K$ satisfying $-y \in K$. A cone that is not acute contains at least one line passing through the origin (together with the origin or without it).

The set L of all solutions (vectors $x \in R^m$) of a homogeneous linear inequality $\langle c, x \rangle = c_1 x_1 + c_2 x_2 + \ldots + c_m x_m \geqq 0$, where c is a fixed nonzero vector from space R^m, represents some convex cone (known as the *closed half-space*).

□

Really, from $\langle c, x \rangle \geqq 0$ it follows that $\alpha\langle c, x \rangle = \langle c, \alpha x \rangle \geqq 0$ for any positive factor α. Hence, L is a cone. Make sure that this is a convex cone. To this end, take two arbitrary points x' and x'' of the cone L. They satisfy the inequalities $\langle c, x' \rangle \geqq 0$ and $\langle c, x'' \rangle \geqq 0$. Multiply the first inequality by an arbitrary number $\lambda \in [0, 1]$ and the second one by $(1 - \lambda)$. The termwise addition of the resulting inequalities yields $\lambda\langle c, x' \rangle + (1 - \lambda)\langle c, x'' \rangle = \langle c, \lambda x' + (1 - \lambda)x'' \rangle \geqq 0$, which establishes the convexity of the cone L. ■

Note that the closed half-space is not an acute cone: together with the nonzero vector $\bar{x}$ satisfying the equality $\langle c, \bar{x} \rangle = 0$, it also contains the vector $-\bar{x}$, since multiplication by -1 does not violate the equality.

If a single linear homogeneous inequality is replaced by a finite system of such, then we get the system of linear homogeneous inequalities. This system also has a convex cone as the solutions set representing the intersection of a finite number of closed half-spaces (the so-called *polyhedral cone*). In the general case, this cone is not acute.

Consider a given collection of vectors $a^1, a^2, \ldots, a^p \in R^m$. It is easy to verify that the aggregate of all nonnegative linear combinations of these vectors (i.e., all vectors of the form $\lambda_1 a^1 + \lambda_2 a^2 + \ldots + \lambda_p a^p$ with nonnegative coefficients $\lambda_1, \lambda_2, \ldots, \lambda_p$) forms some convex *finitely generated* cone K in space R^m. In this case, we say that the collection of vectors $a^1, a^2, \ldots, a^p$ *generates* the convex cone K and write $K = \text{cone}\{a^1, a^2, \ldots, a^p\}$. The vertex belongs to this cone. According to duality theory in convex analysis (see [57], [62]), any finitely generated cone can be represented as the intersection of a finite number of closed half-spaces, i.e., being a polyhedral cone.

The vectors of a convex cone that are not representable as the linear combination of two other vectors of this cone with positive coefficients are called the *edges* or *generators* of the cone. As is well-known [57], any acute polyhedral cone not coinciding with the origin is generated by its edges.

If an acute polyhedral cone is the solution set for some system of linear homogeneous inequalities then the edges of this cone form a *fundamental system of solutions*. And arbitrary solution for this system of linear homogeneous inequalities can be represented as a linear combination of the fundamental system with non-negative coefficients. The fundamental system of solutions can be in principle obtained by exhaustion: just consider all possible subsystems of a definite number of the linear equations resulting from the original system of linear inequalities where inequality signs are replaced by equality ones.

The nonnegative orthant R^m_+ of space R^m, i.e.

$$R^m_+ = \{y \in R^m | y \geq 0_m\},$$

is a convex acute cone (without the vertex) generated by the unit vectors of this space. In the two-dimensional case ($m = 2$), this orthant has the form of the right angle coinciding with the first quarter (see Fig. 2.1). It is generated by the unit

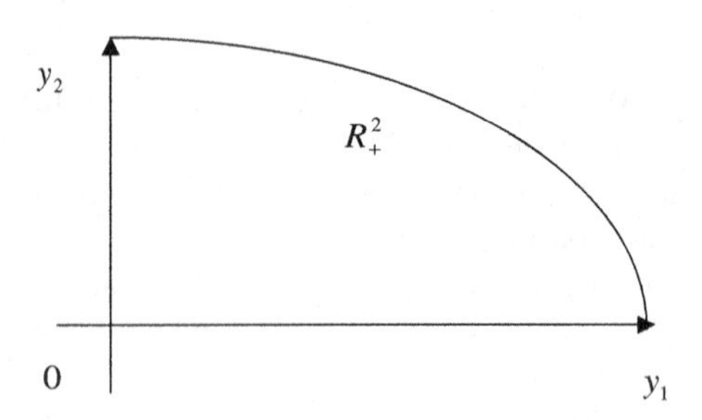

Fig. 2.1 The nonnegative orthant R^2_+

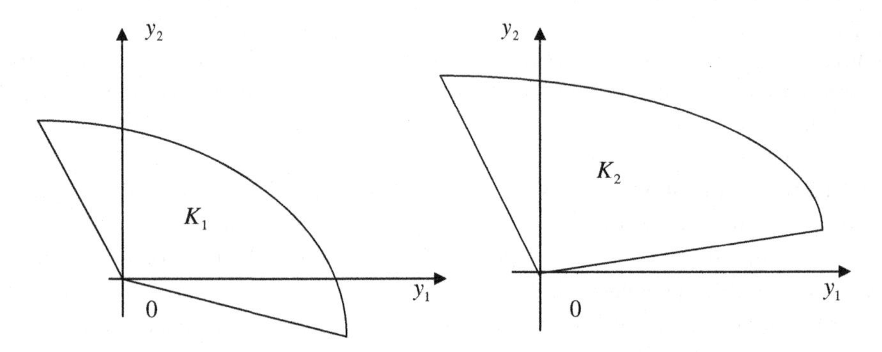

Fig. 2.2 Acute cones K_1 and K_2.

vectors $e^1 = (1,\ 0)$ and $e^2 = (0,\ 1)$, being the intersection of the right and upper closed half-planes (except the origin).

Other examples of acute plane cones, K_1 and K_2, can be observed in Fig. 2.2.

The upper half-plane represents a closed half-space, i.e., a convex cone that is not acute. The convex sets and cones are considered in more detail in [57, 62].

Definition 2.1 A binary relation $\Re$ defined on space R^m, i.e.$\Re \subset R^m \times R^m$, is called a *cone relation* if there exists a cone $K,\ K \subset R^m$, such that for any vectors $y',\ y'' \in R^m$ we have the equivalence

$$y'\Re y'' \Leftrightarrow y' - y'' \in K.$$

Often the right-hand side of the equivalence relationship is written in the form $y' \in y'' + K$ (see Fig. 2.3).

The inequality relations $>$ and $\geq$ considered on space R^m represent some cone relations with the cones $R^m_> = \{y \in R^m | y > 0_m\}$ and R^m_+, respectively.

It appears that any binary relation satisfying Axioms 2 and 4 is a cone relation. This follows from the result below.

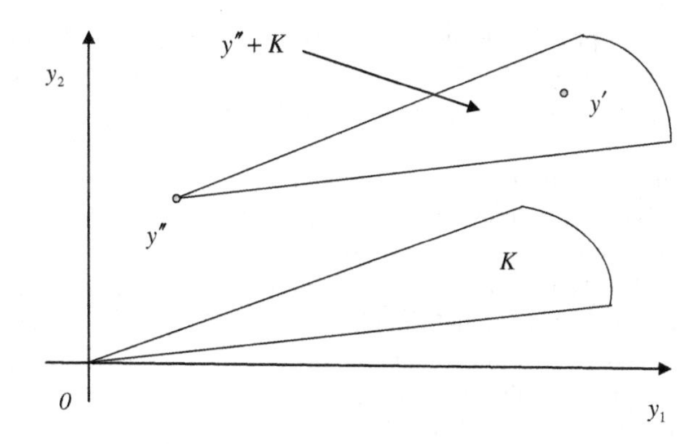

Fig. 2.3 Cone K and its translation $y'' + K$

Lemma 2.2 *Any binary relation* $\Re$ *defined on space* R^m *that has irreflexivity, transitivity and invariance with respect to a linear positive transformation is a cone relation with an acute convex cone not containing the origin. Conversely, each cone relation with the described cone represents a relation defined on* R^m *that has irreflexivity, transitivity and invariance with respect to a linear positive transformation.*

▢ Let $\Re$ be a binary relation defined on R^m that has irreflexivity, transitivity and invariance with respect to a linear positive transformation. Prove that $\Re$ represents a cone relation. To this end, introduce the set

$$K = \{y \in R^m |\, y \Re 0_m\}.$$

Owing to the homogeneity of the relation $\Re$, the set K is a cone. Moreover, for an arbitrary pair of vectors $y', y'' \in R^m$, by additivity we have

$$y' \Re y'' \Leftrightarrow (y' - y'') \Re 0_m \Leftrightarrow (y' - y'') \in K.$$

Therefore, the relation $\Re$ is actually a cone relation with the cone K. Now, it is necessary to verify that the cone K is convex, acute and does not contain the origin.

If $0_m \in K$, then $0_m \Re 0_m$ holds by the definition of the cone K. But this is inconsistent with the irreflexivity of the relation $\Re$. Hence, the cone K does not contain the origin.

To argue the convexity of the cone K, let us choose two arbitrary vectors $y', y'' \in K$ and a number $\alpha \in (0,\ 1)$(note that the values $\alpha = 1$ and $\alpha = 0$ can be omitted from further verification). Owing to the homogeneity of the relation $\Re$, the relationships $y' \Re 0_m$ and $y'' \Re 0_m$ imply $\alpha y' \Re 0_m$ and $(1 - \alpha)y'' \Re 0_m$, respectively. By additivity, the first relationship yields $(\alpha y' + (1 - \alpha)y'') \Re (1 - \alpha)y''$. Now, based on the transitivity of $\Re$, the second and the last relationships give $(\alpha y' + (1 - \alpha)y'') \Re 0_m$, or $(\alpha y' + (1 - \alpha)y'') \in K$, which establishes the convexity of the cone K.

To prove that the cone K is acute, conjecture the opposite. Let there exists a nonzero vector $y \in K$ satisfying the relationship $-y \in K$. For this vector, we have $y \mathfrak{R} 0_m$ and $-y \mathfrak{R} 0_m$. Hence, $(y - y) \mathfrak{R} (-y) \mathfrak{R} 0_m$ by the additive property of $\mathfrak{R}$. Owing to the latter's transitivity, this leads to the relationship $0_m \mathfrak{R} 0_m$ contradicting the irreflexivity of the relation $\mathfrak{R}$.

Now, prove the converse statement. Let $\mathfrak{R}$ be an arbitrary cone relation with an acute convex cone K not containing the origin.

Verify that this relation is irreflexive, transitive and invariant with respect to a linear positive transformation. First, this relation is actually irreflexive (otherwise, the cone K would contain the origin). Second, we verify its transitivity. To this end, select an arbitrary triplet of vectors $y', y'', y''' \in R^m$ satisfying the relationships $y' \mathfrak{R} y''$ and $y'' \mathfrak{R} y'''$. The last two relationships can be rewritten in the form $y' - y'' \in K$ and $y'' - y''' \in K$, whence it follows that there are two definite elements of the cone K. Since the sum of any two elements of a convex cone belongs to this cone, the last relationships yield $y' - y''' \in K$ or, equivalently, $y' \mathfrak{R} y'''$. This result testifies to the transitive property of the relation $\mathfrak{R}$.

And finally, the invariance of the relation $\mathfrak{R}$ follows from the relationships

$$y' \mathfrak{R} y'' \Leftrightarrow y' - y'' \in K \Leftrightarrow (y' + c) - (y'' + c) \in K \Leftrightarrow (y' + c) \mathfrak{R} (y'' + c),$$

$$y' \mathfrak{R} y'' \Leftrightarrow y' - y'' \in K \Leftrightarrow \alpha(y' - y'') \in K \Leftrightarrow \alpha y' - \alpha y'' \in K \Leftrightarrow \alpha y' \mathfrak{R} \alpha y'',$$

which hold for all vectors $c \in R^m$ and any positive number α. ■

Theorem 2.1 *Any binary relation* $\succ$ satisfying Axioms 2, 3 *and* 4 *is a cone relation with an acute convex cone containing the nonnegative orthant* R^m_+ *except the origin. Conversely, each cone relation with the described cone satisfies Axioms* 2, 3 *and* 4.

□ The binary relation $\succ$ satisfying Axioms 2–4 is irreflexive, transitive and invariant with respect to a linear positive transformation.

Necessity. Based on Lemma 2.2, it remains to show that the cone K of the binary relation $\succ$ includes the nonnegative orthant. By Lemma 1.3 from Sect. 1, the Pareto axiom (in terms of vectors) holds, i.e.,

$$y' \geq y'' \Rightarrow y' \succ y''.$$

Rewrite this axiom as the implication

$$y' - y'' \in R^m_+ \Rightarrow y' - y'' \in K.$$

The difference $y' - y''$ can be any vector of the nonnegative orthant R^m_+, and so the above implication means the inclusion $R^m_+ \subset K$.

Sufficiency. If a cone relation is generated by an acute convex cone (without the origin), then by Lemma 2.2 the corresponding cone relation is irreflexive, transitive and invariant with respect to a linear positive transformation (Axioms 2 and 4 are satisfied). On the other hand, this cone contains the nonnegative orthant R_+^m, and therefore the corresponding cone relation also satisfies the Pareto axiom. Obviously, the Pareto axiom implies Axiom 3, and the cone relation under consideration satisfies Axioms 2–4. ■

According to Theorem 2.1, the binary relations satisfying Axioms 2–4 (which are assumed true in the sequel) admit a simple geometrical interpretation. Namely, they represent cone relations with acute convex cones except the origin, and also these cones include the nonnegative orthant R_+^m.

Theorem 2.1 makes it possible to involve convex analysis results for Pareto set reduction.

2.2 Definition of Elementary Information Quantum

2.2.1 Original Multicriteria Choice Problem

The subsequent analysis is dedicated to the multicriteria choice problem that includes

- the set of feasible alternatives X,
- the vector criterion $f = (f_1, f_2, \ldots, f_m)$,
- the preference relation $\succ_X$.

Note that many aspects of this problem become simpler if stated and solved in terms of vectors. As mentioned in previous chapter, all results obtained in terms of alternatives can be easily reformulated in terms of vectors and vice versa. Therefore, further exposition will repeatedly address the *multicriteria choice problem in terms of vectors* that includes

- the set of feasible vectors Y, $Y \subset R^m$,
- the preference relation $\succ$ defined on space R^m.

Recall that the set of feasible vectors is defined by the equality

$$Y = f(X) = \{y \in R^m | y = f(x) \text{for some } x \in X\},$$

while the preference relation $\succ$ represents the extension to the whole space R^m of the preference relation $\succ_Y$ naturally connected to the preference relation $\succ_X$ defined on the set of feasible alternatives X.

Throughout the book below, we assume that Axioms 1–4 hold. Within these conditions (see Lemma 1.3), the Pareto axiom is true, which states that (in terms of vectors) any pair of vectors $y', y'' \in R^m$ such that $y' \geq y''$[1] satisfy the relationship $y' \succ y''$, i.e.,

$$y' \geq y'' \Rightarrow y' \succ y''. \tag{2.1}$$

Under the above assumptions, the DM can compare any two vectors y', y'' from the criterion space R^m using the irreflexive and transitive relation $\succ$. And one and only one of the following cases is realized then:

- $y' \succ y''$, i.e., y' is preferable to y'';
- $y'' \succ y'$, i.e., y'' is preferable to y';
- neither the relationship $y' \succ y''$ nor the relationship $y'' \succ y'$ holds.

2.2.2 *Elementary Information Quantum: Motivation*

Introduce the criteria index set

$$I = \{1, 2, \ldots, m\},$$

and consider the simplest choice problem with two vectors $y', y'' \in R^m$ and the minimum number of different components.

If the vectors y' and y'' have only one different component, e.g., $y'_i \neq y''_i$ and $y'_s = y''_s$ for all $s \in I \backslash \{i\}$, then the relationship $y' \geq y''$ or $y'' \geq y'$ holds. Hence, by Axiom 3, we have $y' \succ y''$ or $y'' \succ y'$, respectively. Therefore, in the elementary case considered, the choice from the two vectors is determined by Axiom 3.

Now, suppose that the vectors y' and y'' have two different components, i.e.,

$$y'_i \neq y''_i, y'_j \neq y''_j; \quad y'_s = y''_s \quad \text{for all } s \in I \backslash \{i, j\},$$

and the equalities $y'_i = y'_j$, $y''_i = y''_j$ do not hold simultaneously. Then one and only one of the following four cases is realized:

$$\begin{aligned} &(1) \quad y'_i > y''_i, \quad y'_j > y''_j; \quad (2) \quad y''_i > y'_i, \quad y''_j > y'_j; \\ &(3) \quad y'_i > y''_i, \quad y''_j > y'_j; \quad (4) \quad y''_i > y'_i, \quad y'_j > y''_j. \end{aligned}$$

Assume that the DM chooses one of these two vectors, i.e., either the relationship $y' \succ y''$ or the relationship $y'' \succ y'$ takes place. Without loss of generality owing to clear symmetry, we believe that the first relationship $y' \succ y''$ is true. And the following question arises immediately. How can the DM's choice be explained?

[1]Recall that the inequality $y' \geq y''$ means $y' \geqq y''$ and $y' \neq y''$.

If the first case from the four ones above is realized, the relationship $y' \succ y''$ results from the Pareto axiom. The second case is impossible: otherwise, by the Pareto axiom, we have the relationship $y'' \succ y'$ that is inconsistent with the relationship $y' \succ y''$ due to the asymmetry of $\succ$.

Now, examine the last two cases. As they are symmetric, it suffices to consider one of them, e.g., the third case. The inequality $y'_i > y''_i$ means that, in terms of criterion i, the vector y' is preferable to y'' for the DM. On the other hand, in terms of criterion j, the vector y'' is preferable to the vector y' since $y''_j > y'_j$. In the final analysis, we have two mutually contradicting conditions and the question is: why does the DM choose the vector y' between the vectors y' and y'' under the existing contradictions? What is the reason of such choice?

Apparently, the most rational explanation for this fact consists in the following. In the contradictory case, the DM is willing to compromise, losing in terms of criterion j for gaining in terms of more important criterion i.

2.2.3 Definition of Elementary Information Quantum

The above arguments applying to the simplest choice problem from an arbitrary pair of vectors motivate the following definition.

Definition 2.2 Let $i, j \in I, i \neq j$. We say that there is *an elementary information quantum about the DM's preference relation with given positive parameters* w_i^*, w_j^* if, for all vectors $y', y'' \in R^m$ such that

$$y'_i - y''_i = w_i^*, y''_j - y'_j = w_j^*, y'_s = y''_s \quad \text{for all } s \in I \backslash \{i, j\}, \tag{2.2}$$

the relationship $y' \succ y''$ holds. Also in this case we say that *criterion* f_i *is more important than criterion* f_j *with parameters* w_i^*, w_j^*.

Remark 2.1 This definition is invariant with respect to the multiplication of the parameters by arbitrary positive number. More specifically, due to the homogeneity of the relation $\succ$, the specification of an elementary information quantum with parameters w_i^*, w_j^* actually generates a similar quantum with the parameters $\alpha \cdot w_i^*$, $\alpha \cdot w_j^*$ for any positive α.

Given an elementary information quantum, the DM that chooses from a pair of vectors (2.2) is willing to sacrifice the quantity w_j^* in terms of criterion f_j for gaining the quantity w_i^* in terms of criterion f_i (the values of all other criteria are fixed).

And the correlation between the quantities w_i^* and w_j^* gives a quantitative estimation for the degree of compromise. For instance, it is possible to consider the ratio w_j^*/w_i^* taking any positive values. However, a more convenient approach is to operate a normalized value from 0 to 1. Apply the transformation $y = x/(1+x)$ to this ratio, arriving at the following notion.

Definition 2.3 Let $i, j \in I, i \neq j$, and there is an elementary information quantum with positive parameters w_i^* and w_j^*. In this case, the number

$$\theta_{ij} = \frac{w_j^*}{w_i^* + w_j^*} = \frac{1}{w_i^*/w_j^* + 1} \in (0, 1)$$

will be called the DM's *coefficient* (or *degree*) *of compromise* for this pair of criteria.

This coefficient shows the share of loss in terms of criterion j the DM accepts against the sum of loss and gain in terms of criterion i. If the coefficient θ_{ij} is close to 1, then the DM incurs a sufficiently large loss in terms of criterion j for obtaining a relatively small gain in terms of criterion i. In this situation, the criterion i has a high importance relatively to the criterion j. Whenever this coefficient is close to 0, the DM is willing to lose in terms of criterion j only for gaining much in terms of the more important criterion. In other words, the degree of importance of criterion i against criterion j is relatively small; and this state of things corresponds to the small degree of compromise. If $\theta_{ij} = 1/2$, then the DM agrees with a definite gain in terms of a more importance criterion at the expense of loss in terms of a less important criterion provided that the loss coincides with the gain.

In addition, take notice that the value of θ_{ij} quantitatively depends on the type of scale used for criteria measurement. For details, we refer to Sect. 2.4.

2.2.4 *Properties of Elementary Information Quantum*

Let us explore the properties of an elementary information quantum.

Theorem 2.2 *If criterion f_i is more important than criterion f_j with given positive parameters w_i^*, w_j^* then criterion f_i is more important than criterion f_j with any pair of positive parameters w_i', w_j' satisfying the inequality $(w_i', -w_j') \geq (w_i^*, -w_j^*)$. In other words, if the DM's degree of compromise is $\theta_{ij} \in (0, 1)$, then this DM possesses any degree of compromise $\theta_{ij}' < \theta_{ij}$.*

□ Choose arbitrarily two positive numbers $w_i', w_j', (w_i', -w_j') \geq (w_i^*, -w_j^*)$, and two vectors $y', y'' \in R^m$ such that

$$y_i' - y_i'' = w_i', y_j'' - y_j' = w_j', y_s' = y_s'' \quad \text{for all } s \in I \backslash \{i, j\}.$$

Prove that $y' \succ y''$.

Consider a vector $z \in R^m$ of the form

$$z_i = y_i'' + w_i^* = y_i' - w_i' + w_i^*,\ z_j = y_j'' - w_j^* = y_j' + w_j' - w_j^*,\ z_s' = y_s'$$
for all $s \in I \backslash \{i, j\}$.

Since $(w'_i, -w'_j) \geq (w^*_i, -w^*_j)$, we have $y' \geq z$. Hence, by the Pareto axiom, $y' \succ z$.

Recall that criterion f_i is more important than criterion f_j with the parameters w^*_i, w^*_j. Using this we get the relationship $z \succ y''$. Together with $y' \succ z$, it leads to the desired result $y' \succ y''$ owing to the transitive property of the relation $\succ$.

Now, prove the second part of the theorem. Let $\theta'_{ij} < \theta_{ij}$. By virtue of Remark 2.1, we may introduce the parameters

$$w'_i = 1 - \theta'_{ij}, w'_j = \theta'_{ij}; \quad w^*_i = 1 - \theta_{ij}, w^*_j = \theta_{ij}.$$

Obviously, for these parameters we have

$$\frac{w'_j}{w'_i + w'_j} = \theta'_{ij}, \quad \frac{w^*_j}{w^*_i + w^*_j} = \theta_{ij}$$

and, in addition, $(w'_i, -w'_j) > (w^*_i, -w^*_j)$.

In this case, using the first part of the theorem (see above), we establish the existence of an elementary information quantum with the parameters w'_i, w'_j, ergo with the degree of compromise θ'_{ij}. ■

The content of Theorem 2.2 well fits the intuitive idea of compromise. In particular, if the DM is willing to lose w^*_j in terms of criterion f_j for gaining w^*_i in terms of criterion f_i, then the DM obviously agrees with a smaller loss w'_j $(w'_j < w^*_j)$ as well as with a greater gain w'_i $(w'_i > w^*_i)$.

Based on the definition of an elementary information quantum and Theorem 2.2, let us analyze the possible cases for an arbitrary pair of different criteria f_i, f_j.

In fact, one and only one of the three cases are possible as follows:

1. At least one positive number from the interval $(0, 1)$ represents the degree of compromise for criteria f_i and f_j, and at least one number does not;
2. None of the positive numbers from the interval $(0, 1)$ is the degree of compromise for criteria f_i, f_j. In this case, we shall say that *criterion f_i is not more important than criterion f_j*;
3. Any positive number from the interval $(0, 1)$ is the degree of compromise for criteria f_i, f_j. In this case, we shall say that *criterion f_i is incomparably more important than criterion f_j*.

Let us investigate the first case. If at least one number $\theta_{ij} \in (0, 1)$ is the degree of compromise, then by Theorem 2.2 any smaller number within this interval is also the degree of compromise for the pair of criteria under consideration. Construct two disjoint sets A and B using the following procedure. Add on to the former set all numbers from the interval $(0, 1)$ that are the degrees of compromise for this pair of criteria; naturally, $A \neq \emptyset$. The latter set B comprises all numbers from the interval that are not the degrees of compromise; by the data, $B \neq \emptyset$. Clearly, the described procedure yields $A \cup B = (0, 1)$, and the inequality $a < b$ holds for all $a \in A$, $b \in B$. This means that the sets A and B define a section of the interval $(0, 1)$. By the Dedekind principle, there exists a unique number $\bar{\theta}_{ij} \in (0, 1)$ implementing this section, further called the *limit degree of compromise*.

Note that the number $\bar{\theta}_{ij}$ may be the degree of compromise or not. In other words, either $\bar{\theta}_{ij} \in A$ or $\bar{\theta}_{ij} \notin A$ holds.

2.2.5 Connection to Lexicographic Relation

The preference relation $\succ$ satisfying Axioms 2–4 and the lexicographic[2] relation have a certain connection revealed by the next statement in terms of an ordered collection of incomparably more important criteria.

Theorem 2.3 *The irreflexive, transitive and invariant relation $\succ$ defined on space R^m is a lexicographic relation if and only if criterion f_1 is incomparably more important than criterion f_2, criterion f_2 is incomparably more important than criterion f_3,..., criterion f_{m-1} is incomparably more important than criterion f_m.*

□ Necessity. Let the relation $\succ$ be lexicographic. In this case, for arbitrary vectors $y', y'' \in R^m$, we have the logical propositions

$$(1) \quad y'_1 > y''_1 \quad \Rightarrow \quad y' \succ y'',$$

$$(2) \quad y'_1 = y''_1, y'_2 > y''_2 \quad \Rightarrow \quad y' \succ y'',$$

$$(3) \quad y'_1 = y''_1, y'_2 = y''_2, y'_3 > y''_3 \quad \Rightarrow \quad y' \succ y''$$

$$\cdots\cdots\cdots\cdots\cdots\cdots\cdots\cdots\cdots$$

$$m) \quad y'_i = y''_i, i = 1, 2, \ldots, m-1; \quad y'_m > y''_m \quad \Rightarrow \quad y' \succ y''.$$

The first proposition implies the relationship $y' \succ y''$ for two arbitrary vectors $y', y'' \in R^m$ satisfying $y'_1 > y''_1, y'_2 < y''_2, y'_3 = y''_3, \ldots, y'_m = y''_m$. This means that criterion f_1 is incomparably more important than criterion f_2.

Similarly, using the second proposition, we conclude that criterion f_2 is incomparably more important than criterion f_3, and so on; in the final analysis, the incomparably higher importance of criterion f_{m-1} against criterion f_m follows from the $(m-1)$-th proposition.

Sufficiency.[3] For each $i = 1, 2, \ldots, m-1$, let criterion f_i be incomparably more important than criterion f_{i+1}. Prove that the relation $\succ$ is lexicographic.

---

[2]The definition of a lexicographic relation can be found in Sect. 1.2.

[3]The proof is suggested by O.V. Baskov.

Consider two arbitrary vectors y', $y'' \in R^m$. If they coincide, then none of them is lexicographically greater than the other, which agrees with the definition of a lexicographic relation.

Let $y' \neq y''$. Denote by i the minimum index such that $y'_i \neq y''_i$. Without loss of generality, assume that $y'_1 < y''_1$. It is required to show that $y'' \succ y'$. The proof has the form of an algorithm with the following steps.

Step 1. Compare the numbers y'_m and y''_m. If $y'_m = y''_m$, proceed to Step 2 by setting $z^1 = y'$. If $y'_m < y''_m$, introduce the vector $z^1 = (y'_1, \ldots, y'_{m-1}, y''_m)$, which satisfies the relationship $z^1 \succ y'$ due to the compatibility axiom. Then pass to Step 2.
If $y'_m > y''_m$, fix an arbitrary $\alpha > y'_{m-1}$ and introduce the vector $z^1 = (y'_1, \ldots, y'_{m-2}, \alpha, y''_m)$. Since the criterion f_{m-1} is incomparably more important than the criterion f_m, we obtain $z^1 \succ y'$. Next, move to Step 2.

Step 2. By analogy, continue the comparison of z^1_{m-1} and y''_{m-1}. And so on.

Step k+1. At this step, we have $z^k_j = y''_j$, $j = i+2, \ldots, m$. Compare z^k_{i+1} and y''_{i+1}. If $z^k_{i+1} = y''_{i+1}$, then the compatibility axiom dictates that $y'' \succ z^k$. In the case $z^k_{i+1} < y''_{i+1}$, we get $y'' \geq z^k$. Owing to the Pareto axiom, hence it appears that $y'' \succ z^k$. If $z^k_{i+1} > y''_{i+1}$, then $y'_i = z^k_i < y''_i$ and the incomparably higher importance of the criterion f_i against the criterion f_{i+1} again give $y'' \succ z^k$.

As a result, we arrive at the chain of relationships $y'' \succ z^k \succ \ldots z^1 \succ y'$ where (at some but not all positions) the preference symbol $\succ$ can be replaced by the equality sign. In combination with the transitivity of the preference relation, this leads to the desired relationship $y'' \succ y'$. ■

2.3 Pareto Set Reduction Using Elementary Information Quantum

2.3.1 Simplification of Basic Definition

Definition 2.2 reveals the whole essence of an elementary information quantum about the DM's preference relation. This definition involves two numerical parameters used to measure the degree of compromise.

To verify that criterion f_i is more important than criterion f_j, by Definition 2.2 we have to compare infinitely many pairs of vectors y', $y'' \in R^m$ such that

$$y'_i - y''_i = w^*_i > 0, y''_j - y'_j = w^*_j > 0, y'_s = y''_s \quad \text{for all } s \in I \backslash \{i, j\}. \tag{2.3}$$

And if for any pair above, the first vector y' every time appears preferable to the second one y'', then by Definition 2.2 there is a given elementary information quantum with the corresponding parameters.

It is absolutely clear that such verification appears non-implementable in practice due to infinitely many pairs of vectors for comparison. Actually, this verification is not required, as the preference relation possesses invariance. The whole procedure can be reduced to comparing merely a pair of vectors y', y'' satisfying (2.3). The following result gives the details.

Theorem 2.4 *In Definition* 2.2, *the vectors* y', y'' *can be assumed fixed. Particularly,*

$$y'_i = w^*_i, y'_j = -w^*_j \text{ and } y'_s = 0 \quad \text{for all } s \in I \backslash \{i,j\}, y'' = 0_m, \tag{2.4}$$

or

$$y'_i = 1 - \theta_{ij},\ y'_j = -\theta_{ij} \text{ and } y'_s = 0 \quad \text{for all } s \in I \backslash \{i,j\}, y'' = 0_m, \tag{2.4'}$$

where θ_{ij} *is the degree of compromise.*

☐ Consider two arbitrary vectors y' and y'' satisfying (2.3). Obviously,

$$y'_i > y''_i \Leftrightarrow y'_i - y''_i > 0,$$
$$y''_j > y'_j \Leftrightarrow y''_j - y'_j > 0.$$

Denote $\bar{y}_i = y'_i - y''_i = w^*_i$, $\bar{y}_j = y'_j - y''_j = -w^*_j$, where $\bar{y}_s = 0$ for all $s \in I \backslash \{i,j\}$. By the additivity of the preference relation $\succ$, we have

$$y' \succ y'' \Leftrightarrow (y' - y'') \succ 0_m \Leftrightarrow \bar{y} \succ 0_m,$$

where the vector $\bar{y}$ has only two nonzero components, namely, components i and j being $\bar{y}_i$ and $\bar{y}_j$, respectively. This means that Definition 2.2 in the general form is equivalent to itself in the "simplified" form with the fixed vectors $y' = \bar{y}$ and $y'' = 0_m$.

Hence, in Definition 2.2 the vectors y', y'' can be assumed fixed.

Now, we prove the remainder of Theorem 2.4. The relationship $\bar{y} \succ 0_m$ for the above vector $\bar{y}$ is equivalent to the relationship $\alpha \bar{y} \succ 0_m$ with any positive number α by the homogeneity of the preference relation $\succ$. Choosing $\alpha = -\theta_{ij}/\bar{y}_j$ and setting $\hat{y} = \alpha \bar{y}$ yield

$$\hat{y}_i = \alpha\bar{y}_i = -\frac{\theta_{ij}\bar{y}_i}{\bar{y}_j} = \frac{\theta_{ij}w_i^*}{w_j^*} = \frac{w_i^*}{w_i^* + w_j^*} = 1 - \theta_{ij},$$

$$\hat{y}_j = \alpha\bar{y}_j = -\frac{\theta_{ij}\bar{y}_j}{\bar{y}_j} = -\theta_{ij},$$

$$\hat{y}_s = \alpha\bar{y}_s = \alpha 0 = 0 \quad \text{for all } s \in I\backslash\{i,j\}.$$

Therefore, the relationship $\bar{y} \succ 0_m$ is equivalent to the relationship $\hat{y} \succ 0_m$, where the vector $\hat{y}$ has the same components

$$\hat{y}_i = 1 - \theta_{ij}, \hat{y}_j = -\theta_{ij}; \hat{y}_s = 0 \quad \text{for all } s \in I\backslash\{i,j\},$$

as the vector y' from (2.4′). ■

According to the aforesaid, the preference relation $\succ$ is supposed invariant with respect to a linear positive transformation. Using Theorem 2.4, we introduce a new (simplified) definition of an elementary information quantum.

Definition 2.4. Let $i,j \in I$, $i \neq j$. We say that there is a *given elementary information quantum with positive parameters* w_i^*, w_j^* *(with the degree of compromise* $\theta_{ij} \in (0,1)$) if the relationship $y' \succ 0_m$ holds for the vector $y' \in R^m$ of form (2.4) (form (2.4′), respectively).

To verify that criterion f_i is more important than criterion f_j with the degree of compromise $\theta_{ij} \in (0,1)$, by Definition 2.4 it suffices to check that the vector y' of form (2.4) is preferable to the zero vector, i.e. $y' \succ 0_m$. For instance, if the vector $(0.7, -0.3, 0)$ appears preferable to $(0, 0, 0)$ for the DM, then the first criterion is more important for the DM than the second criterion with the degree of compromise $\theta_{12} = 0.3$.

2.3.2 *Pareto Set Reduction Based on Elementary Information Quantum*

The next result shows how the available information about the preference relation in the form of an elementary quantum can be used for reducing the search space of selectable vectors.

Theorem 2.5 (in terms of vectors). *Assume that there exists an elementary information quantum with positive parameters* w_i^* *and* w_j^* *(with the degree of compromise* $\theta_{ij} \in (0,1)$). *Then for any set of selectable vectors* $C(Y)$ *we have*

$$C(Y) \subset \hat{P}(Y) \subset P(Y), \tag{2.5}$$

where $\hat{P}(Y)$ *is the set of feasible vectors corresponding to the set of Pareto optimal alternatives in the multicriteria problem with the initial set of feasible alternatives* X *and the "new" vector criterion* $\hat{f} = (\hat{f}_1, \hat{f}_2, \ldots, \hat{f}_m)$ *(i.e.,* $\hat{P}(Y) = f(P_{\hat{f}}(X))$*) with the components calculated by*

$$\hat{f}_j = w_j^* f_i + w_i^* f_j, \hat{f}_s = f_s \quad \text{for all } s \in I \backslash \{j\}, \tag{2.6}$$

or

$$\hat{f}_j = \theta_{ij} f_i + (1 - \theta_{ij}) f_j, \hat{f}_s = f_s \quad \text{for all } s \in I \backslash \{j\}. \tag{2.6'}$$

□ The proof consists of four parts.

I. Denote by K the acute convex cone (without the origin) of the cone preference relation $\succ$. By the hypothesis of Theorem 2.5 and Definition 2.4, the vector y' described by equalities (2.4) satisfies the relationship $y' \succ 0_m$. The latter is equivalent to the inclusion $y' \in K$.

Consider the collection of unit vectors $e^1, e^2, \ldots, e^m$ of space R^m; here component s of the vector e^s is 1 and the other components are 0. Let M be the convex cone (without the origin) generated by the collection of linear independent[4] vectors

$$e^1, \ldots, e^{i-1}, y', e^{i+1}, \ldots, e^m. \tag{2.7}$$

The cone M coincides with the set of all vectors representable as the linear combinations

$$\lambda_1 e^1 + \ldots + \lambda_{i-1} e^{i-1} + \lambda_i y' + \lambda_{i+1} e^{i+1} + \ldots + \lambda_m e^m$$

of the vectors from collection (2.7) with the nonnegative coefficients $\lambda_1, \lambda_2, \ldots, \lambda_m$ that are not zero simultaneously.

Check that the cone M is acute. If not, there exists a nonzero vector $y \in M$ such that $-y \in M$. According to the aforesaid,

$$y = \lambda_1 e^1 + \ldots + \lambda_{i-1} e^{i-1} + \lambda_i y' + \lambda_{i+1} e^{i+1} + \ldots + \lambda_m e^m,$$
$$-y = \lambda'_1 e^1 + \ldots + \lambda'_{i-1} e^{i-1} + \lambda'_i y' + \lambda'_{i+1} e^{i+1} + \ldots + \lambda'_m e^m,$$

where all coefficients of the linear combinations are nonnegative and each of the collections $\lambda_1, \lambda_2, \ldots, \lambda_m$ and $\lambda'_1, \lambda'_2, \ldots, \lambda'_m$ is not zero simultaneously. The sum of

[4]Indeed, vectors (2.7) form a linear independent system, since the matrix composed of them has rank m.

two elements of a cone belongs to this cone; hence, by summing up the last two equalities, we obtain

$$0_m = (\lambda_1 + \lambda_1')e^1 + \ldots + (\lambda_{i-1} + \lambda_{i-1}'')e^{i-1} + (\lambda_i + \lambda_i')y' + (\lambda_{i+1} + \lambda_{i+1}')e^{i+1} + \ldots + (\lambda_m + \lambda_m')e^m,$$

where at least one coefficient of the linear combination in parentheses is nonzero. However, owing to the linear independence of vectors (2.7), the last equality implies that all coefficients of the linear combination are zero. This contradiction to the initial hypothesis testifies that the cone M is acute.

II. Now, demonstrate that the cone M coincides with the set of all nonzero solutions to the following system of linear homogeneous inequalities:

$$\begin{aligned} y_s &\geqq 0 \qquad \text{for all } s \in I\backslash\{j\}, \\ w_j^* y_i + w_i^* y_j &\geqq 0. \end{aligned} \tag{2.8}$$

To this end, find the fundamental system of solutions for the system of inequalities (2.8) and make sure that it coincides with collection (2.7).

For obtaining the fundamental system of solutions for the system of inequalities (2.8), consider the corresponding collection of linear equations

$$\begin{aligned} y_s &= 0 \quad \text{for all } s \in I\backslash\{j\}, \\ w_j^* y_i + w_i^* y_j &= 0, \end{aligned} \tag{2.9}$$

which can be rewritten as[5]

$$\begin{aligned} \langle e^s, y\rangle &= 0 \quad \text{for all } s \in I\backslash\{j\}, \\ \langle \tilde{y}, y\rangle &= 0, \end{aligned} \tag{2.10}$$

where $\tilde{y} = (\tilde{y}_1, \tilde{y}_2, \ldots, \tilde{y}_m)$ and

$$\tilde{y}_i = w_j^*, \tilde{y}_j = w_i^*, \tilde{y}_s = 0 \quad \text{for all } s \in I\backslash\{i, j\}.$$

The number of equations in (2.10) is m. An arbitrary collection of $m-1$ vectors obtained from $e^1, \ldots, e^{j-1}, \tilde{y}, e^{j+1}, \ldots, e^m$ by removing a single vector appears linearly independent. And so, to find the fundamental system of solutions for the system of inequalities (2.8), it suffices to go over the nonzero solutions to each subsystem constructed from $m-1$ equalities of the original system (2.10). Among them, one should choose the vectors satisfying the system of inequalities (2.8).

[5]Recall that, for m-dimensional vectors a and b, the notation $\langle a, b\rangle$ gives their *scalar product*: $\langle a, b\rangle = \sum_{i=1}^{m} a_i b_i$.

We will sequentially eliminate one equation from system (2.10), seeking for the nonzero solutions to the resulting "truncated" system. With the last equation eliminated from (2.10), e.g., the vector e^j is a nonzero solution to the "truncated" system. After elimination of the equation $\langle e^s, y\rangle = 0$ (where $s \neq i$), the vector e^s can be chosen as a nonzero solution to the "truncated" system. As easily verified, the "truncated" system without the equation $\langle e^i, y\rangle = 0$ has the nonzero solution y'. This procedure yields the fundamental system of solutions $e^1, \ldots, e^{i-1}, y', e^{i+1}, \ldots, e^m$ to the system of inequalities (2.8). The fundamental system coincides with the vector collection (2.7) generating the cone M of the cone preference relation $\succ$. Therefore, the cone M represents the set of nonzero solutions to the system of linear inequalities (2.8).

III. As mentioned in the beginning of the proof, the inclusion $y' \in K$ takes place. By Theorem 2.1, we have the relationship $R^m_+ \subset K$. The cone R^m_+ is generated by the collection of unit vectors $e^1, e^2, \ldots, e^m$. Since K represents a convex cone, together with vectors (2.7) it surely contains all nonzero linear combinations of vectors (2.7) with nonnegative coefficients, i.e., $M \subset K$. Finally, we get the inclusions

$$R^m_+ \subset M \subset K,$$

yielding

$$\text{Ndom } Y \subset \hat{P}(Y) \subset P(Y), \tag{2.11}$$

where

$$P(\hat{Y}) = \{y^* \in Y|\ \text{there exists no } y \in Y \text{ such that } y - y^* \in M\}$$

is the set of nondominated elements of the set Y with respect to the cone relation with the cone M.

IV. Choose arbitrarily two elements $x, x^* \in X, y = f(x), y^* = f(x^*)$ such that $f(x) \neq f(x^*)$. As shown in part II, the cone M coincides with the set of nonzero solutions to the system of linear inequalities (2.8), and hence the inclusion $f(x) - f(x^*) \in M$ takes place if and only if the vector $y = f(x) - f(x^*)$ is a nonzero solution to (2.8), i.e.,

$$\begin{pmatrix} f_1(x) - f_1(x^*) \\ \cdot \quad \cdot \quad \cdot \quad \cdot \quad \cdot \quad \cdot \quad \cdot \\ f_{j-1}(x) - f_{j-1}(x^*) \\ w_j^*(f_i(x) - f_i(x^*)) + w_i^*(f_j(x) - f_j(x^*)) \\ f_{j+1}(x) - f_{j+1}(x^*) \\ \cdot \quad \cdot \quad \cdot \quad \cdot \quad \cdot \quad \cdot \quad \cdot \\ f_m(x) - f_m(x^*) \end{pmatrix} \geq 0_m.$$

The last inequality can be rewritten in the compact form $\hat{f}(x) - \hat{f}(x^*) \in R^m_+$, or $\hat{f}(x) \geq \hat{f}(x^*)$ where $\hat{f}$ is defined by (2.6). And therefore the relationship $y - y^* \in M$ for the vectors $y = f(x), y^* = f(x^*)$ is equivalent to the inequality $\hat{f}(x) \geq \hat{f}(x^*)$. Subsequently, $\hat{P}(Y) = f(P_{\hat{f}}(X))$.

By the hypothesis of the current theorem and Lemma 1.2, we have the inclusion $C(Y) \subset$ Ndom Y for arbitrary set $C(Y)$. And so, inclusions (2.11) lead to inclusions (2.5), which were to be established.

The vector criterion (2.6′) is obtained from (2.6) by dividing component j of the latter by the positive number $w_i^* + w_j^*$. Such an operation clearly do not modify the Pareto set $\hat{P}(Y)$. ■

According to the Edgeworth-Pareto principle, all selectable vectors must belong to the Pareto set. If the multicriteria choice problem includes additional information about the DM's willingness to compromise while comparing the values of two certain criteria, then Theorem 2.5 serves for Pareto set reduction based on this information without losing any selectable vectors. In other words, some vectors can be eliminated from the Pareto set, since they would not be selected for sure.

For justice' sake, we have to note the following. In definite cases (especially if the degree of compromise is close to 0, viz. the criteria f_j and $\hat{f}_j$ almost coincide), the reduction may fail due to the identical Pareto sets in terms of the "old" and "new" vector criteria, i.e. $\hat{P}(Y) = P(Y)$. One can say that in such cases the available information about the preference relation is not rich in content.

Theorem 2.4 acquires the following form in terms of alternatives.

Theorem 2.6 (in terms of alternatives). *Assume that criterion f_i is more important than criterion f_j with given positive parameters w_i^*, w_j^* (with the degree of compromise $\theta_{ij} \in (0, 1)$). Then for any set of selectable alternatives $C(X)$ we have*

$$C(X) \subset P_{\hat{f}}(X) \subset P_f(X), \tag{2.12}$$

where $P_{\hat{f}}(X)$ is the set of Pareto optimal alternatives in the multicriteria problem with the set of feasible alternatives X and the "new" vector criterion $\hat{f} = (\hat{f}_1, \hat{f}_2, \ldots, \hat{f}_m)$ with the components calculated by formulas (2.6) *or* (2.6′).

Figure 2.4 illustrates the inclusions (2.12).

Commenting on Theorem 2.5, first of all we emphasize its universalism. Namely, there exist no requirements to the set of feasible alternatives X and the

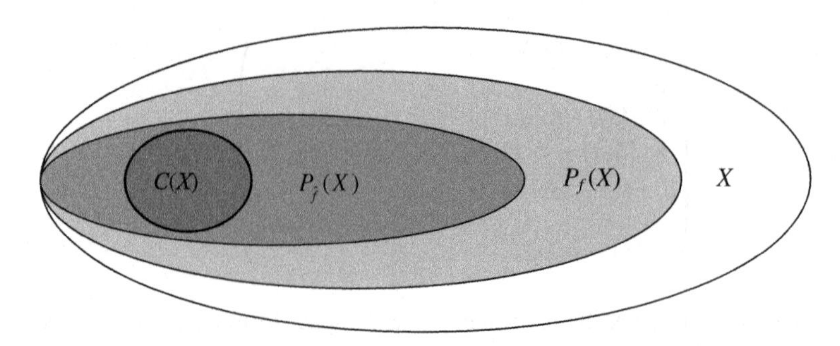

Fig. 2.4 Nested sets

vector criterion f. This theorem is hence applicable to any multicriteria choice problem satisfying Axioms 1–4. And the set of feasible alternatives (and vectors) may be finite or infinite, while the functions $f_1, f_2, \ldots, f_m$ may belong to an arbitrary class (being nonlinear, nonconvex, nonconcave or discontinuous). The only constraint in the conditions of Theorem 2.5 concerns the DM's behavior: during the choice process, the DM must act "reasonably" in the sense that its preference relation necessarily meets Axioms 1–4. Second, the "new" criterion $\hat{f}$ is recalculated using the "old" one f by a very simple formula, see (2.6). According to it, the "new" vector criterion is obtained from the "old" counterpart by replacing the less important criterion f_j for the positive linear combination of the criteria f_i and f_j with the parameters w_i^*, w_j^*. The other "old" criteria remain the same. As easily seen, this "recalculation" of criterion j does not affect many fruitful optimization-oriented properties of the criteria f_i and f_j. For instance, if these criteria are continuous, concave, convex or linear, the new criterion $\hat{f}_j$ inherits the same properties.

The simplest recalculation formula appears in the case of linear criteria. We state the corresponding result below.

Corollary 2.1 *In addition to the hypothesis of Theorem* 2.5, *let* $X \subset R^n$ *and let the criteria* f_i *and* f_j *be linear, i.e.,*

$$f_k(x) = \langle c^k, x \rangle = \sum_{l=1}^{n} c_l^k x_l, \quad k = i, j,$$

where $c^k = (c_1^k, c_2^k, \ldots, c_n^k)$. *Then the new criterion* j *has the form* $\hat{f}_j(x) = \langle \hat{c}, x \rangle$ *with* $\hat{c} = w_j^* c^i + w_i^* c^j$, *or*

$$\hat{c} = \theta_{ij} c^i + (1 - \theta_{ij}) c^j. \tag{2.13}$$

This result immediately follows from formula (2.6) and the linear property of the scalar product of vectors.

Equality (2.13) admits a clear interpretation if the set of feasible alternatives is a subspace in the two-dimensional vector space, i.e., $X \subset R^2$ (see Fig. 2.5).

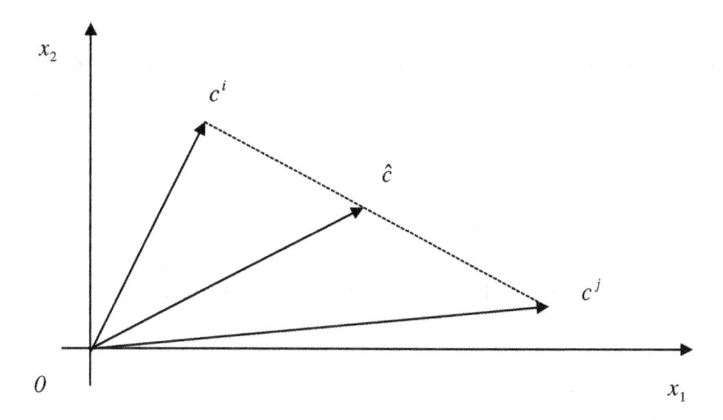

Fig. 2.5 Vectors c^i, c^j and $\hat{c}$

The closer is the degree of compromise θ_{ij} to 0, the closer is the end of the vector $\hat{c}$ to that of the vector c^j. As we increase θ_{ij} within the interval (0, 1), the vector "attracts" the vector $\hat{c}$ associated with the new criterion j. In the case $\theta_{ij} = 0.5$, the end of the vector $\hat{c}$ is in the middle of the line segment connecting the ends of the two vectors c^i and c^j. If the degree of compromise is close to 1, then the vector $\hat{c}$ slightly differs from c^i, which means that the vector criterion $\hat{f}$ includes two almost identical criteria. And the impact of the less important criterion f_j associated with the vector c^j on the solution of the multicriteria choice problem becomes negligible.

2.3.3 Geometrical Aspects

As a rule, the preference relation $\succ$ guiding the DM choice process is not completely defined (i.e., fragmentary) in the multicriteria choice problems. Throughout this book, we assume that it merely satisfies Axioms 1–4. Under these conditions, by Theorem 2.1 the preference relation $\succ$ is a cone relation with an (unknown) acute convex cone K except the origin. Furthermore, the cone K contains the nonnegative orthant, i.e., $R^m_+ \subset K$. This gives the inclusion Ndom $Y \subset P(Y)$, in combination with $C(Y) \subset$ Ndom Y yielding

$$C(Y) \subset P(Y). \tag{2.14}$$

The last inclusion expresses the Edgeworth-Pareto principle, which states that the choice should be performed within the Pareto set. As mentioned in Sect. 1.4, this principle is applicable to any multicriteria choice problem satisfying Axioms 1–3. Here is an alternative formulation of the principle: *the Pareto set represents an upper estimate for the set of selectable vectors.*

Now, suppose that (besides Axioms 1–4 satisfied by the multicriteria choice problem) we have additional information that criterion f_i is more important than criterion f_j with the degree of compromise $\theta_{ij} \in (0, 1)$. In geometrical terms, the existence of such information means the specification of a vector $y' \in R^m$ of form (2.4) with the inclusion $y' \in K$. Consequently, the cone K contains not only the nonnegative orthant, but also the vector y' beyond this orthant.

Consider the cone M coinciding with the set of all nonzero nonnegative linear combinations of the vectors $e^1, \ldots, e^{i-1}, y', e^{i+1}, \ldots, e^m$, see the proof of Theorem 2.5. In the course of this proof, we have also established the inclusions $R^m_+ \subset M \subset K$, where $M \neq R^m_+$. These inclusions imply

$$C(Y) \subset \text{Ndom}\, Y \subset \text{Ndom}_M\, Y \subset P(Y),$$

where

$$\text{Ndom}\, Y = \{y^* \in Y |\ \text{there exists no } y \in Y \text{ such that } y - y^* \in K\},$$

$$\text{Ndom}_M Y = \{y^* \in Y |\ \text{there exists no } y \in Y \text{ such that } y - y^* \in M\},$$

$$P(Y) = \{y^* \in Y |\ \text{there exists no } y \in Y \text{ such that } y - y^* \in R^m_+\}.$$

Hence, the upper estimate (2.14) for the unknown set of selectable vectors is refined to

$$C(Y) \subset \text{Ndom}_M\, Y.$$

Note that, the wider is the cone M in comparison with the nonnegative orthant R^m_+, the narrower is the set $\text{Ndom}_M Y$ in comparison with $P(Y)$.

Thus, using an elementary information quantum, one can extract in the unknown cone K a cone M wider than R^m_+ (see Fig. 2.6), thereby constructing a more precise upper estimate for the set of selectable vectors as against the estimate yielded the Edgeworth-Pareto principle.

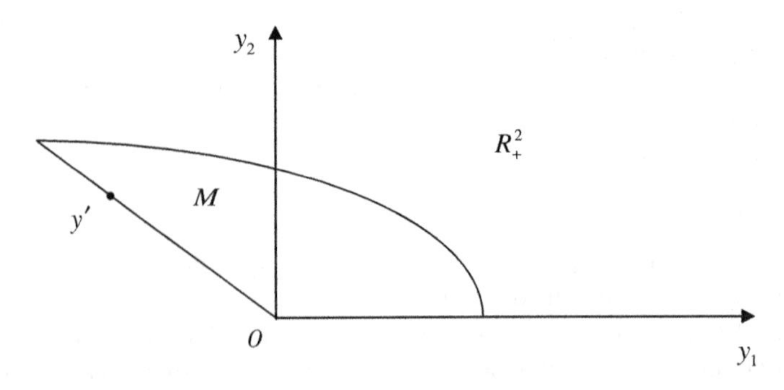

Fig. 2.6 Cones M and R^2_+.

Example 2.1 Let $m = 2$ and $Y = \{y^1, y^2, y^3\}$, where

$$y^1 = (4, 1), y^2 = (3, 2), y^3 = (1, 3).$$

Here all the three feasible vectors are Pareto optimal. In other words, the Edgeworth-Pareto principle does not assist in reducing the search space of selectable vectors.

Imagine that the first criterion is more important than the second one with the degree of compromise 0.5. Geometrically, this means that $y' = (0.5, -0.5) \in K$.

Figure 2.7 shows the three feasible vectors and the cone M translated into the points corresponding to the second and third feasible vectors.

Clearly, neither the second nor third vector can be selected, as both have dominating vectors:

$$y^2 \in y^3 + M, y^1 \in y^2 + M.$$

And the only selectable vector is hence the first one y^1. In other words, if the set of selectable vectors is non-empty in this problem, then it consists of the first vector only.

The same conclusion can be drawn using Theorem 2.5. Really, by formula (2.6) the new criterion 2 acquires the form $0.5y_1 + 0.5y_2$ and, as easily found,

$$\hat{f}(X) = \{(4, 2.5), (3, 2.5), (1, 2)\}.$$

In this set, the first vector is the only Pareto optimal one; it corresponds to the vector y^1. Therefore, this (and only this vector) can appear selectable from Y if the selectable vectors exist.

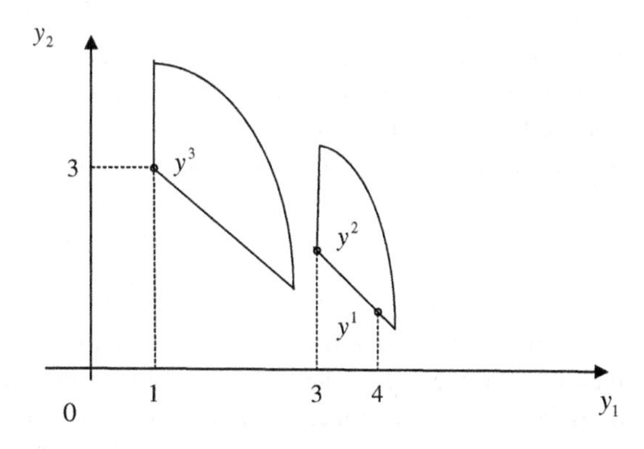

Fig. 2.7 The dominated vectors y^2 and y^3

2.4 Scales of Criteria and Invariance of Measurements

2.4.1 Quantitative and Qualitative Scales

As mentioned earlier, all criteria $f_1, f_2, \ldots, f_m$ in the multicriteria choice problem statement have numerical values. Therefore, the inclusion $y_i = f_i(x) \in R$ holds for any $x \in X$ and each $i = 1, 2, \ldots, m$. This information about the criteria is enough for the rigorous mathematical formulation of the multicriteria choice problem.

However, in real applications the numerical values of the criteria are the measurement results in a certain scale. For instance, if a criterion reflects the value, cost or profit of a project, these quantities can be expressed in RUB, USD, EURO or other monetary units. The lengths of different objects are measured in meters, inches, foots, yards, and so on. Hours, seconds, years, millions of years, etc. are used for time intervals. Consequently, in specific applications the values of criteria are associated with a certain scale, being expressed in definite units of measure.

There exist various measurement scales. Whenever it is required to count the number of objects, people, items, etc., one adopts the so-called *absolute scale*. This scale has a fixed reference point (0) and a fixed spacing (1). Two individuals performing independent measurements of same quantities in the absolute scale (two measurers) must obtain the identical results. In addition, note that this scale has a unique unit of measure for all measurers.

Different units of measure are used for measuring the physical characteristic of mass. As is well-known, the mass of an object can be expressed in kilograms, pounds, tones, poods, etc. Here only the reference point (0) is fixed for all measurers, which corresponds to the absence of mass; and the scale spacing may vary for measurers. Thereby, for a same object, the measurements results y_i' and y_i'' obtained by two measurers in different units of measure differ by some fixed positive factor α_i, i.e., $y_i' = \alpha_i y_i''$. In this case, the measurement results are defined within the transformation $\phi_i(y_i) = \alpha_i y_i$, $\alpha_i > 0$. Such a scale is called the *ratio scale*, which can be explained as follows. Regardless of the unit of measure, the measurements in this scale yield the same ratios for different measurers. Really, assume that, for two objects, measurers 1 and 2 obtain the values y_i', y_i'' and $\tilde{y}_i', \tilde{y}_i''$, respectively. Since $\tilde{y}_i' = \alpha_i y_i'$ and $\tilde{y}_i'' = \alpha_i y_i''$ for some $\alpha_i > 0$, we have the equalities

$$\frac{\tilde{y}_i'}{\tilde{y}_i''} = \frac{\alpha_i y_i'}{\alpha_i y_i''} = \frac{y_i'}{y_i''},$$

which mean that the ratios of the measurements are preserved for two different measurers in the ratio scale. And so, if a measurer concludes that, e.g., the mass of an object is twice as much as that of the other, then another measurer (operating different units of measure) must come to the same conclusion. This testifies that, while comparing the measurement results in the ratio scale, the statement "object 1 is α_i times greater (smaller) than object 2" actually makes sense.

Obviously, such quantities as profit, costs, etc. expressed in currency units should be measured in the ratio scale, too.

Another measurement scale has a given spacing and an unfixed reference point (for different measurers). A possible example is the chronology scale–passing from one chronology to another requires an appropriate variation in the reference point. More precisely, the *difference scale* is a scale in which the measurement results are defined within the transformation $\phi_i(y_i) = y_i + c_i$ with a fixed constant c_i. The measurements in this scale preserve the differences between two different measurements performed by distinct measurers. In other words, for the measurements in the difference scale, a sensible statement has the form "object 1 is greater (smaller) than object 2 by the constant c_i." For example, the reign of Tsar Nicholas II in Russia calculated according to the Gregorian and Julian calendars is the same (as well as in any other calendar).

The interval scale is a scale in which the measurement results are defined within (are invariant with respect to) the linear positive transformation $\phi_i(y_i) = \alpha_i y_i + c_i$, where $\alpha_i > 0$ and c_i represent fixed constants. A typical example of such a scale is a temperature scale. As is well-known, the Celsius scale and the Fahrenheit scale serve for temperature measurements. Transition from one scale to the other employs the formula $\tilde{y}_i = \alpha_i y_i + c_i$.

Each measurer choosing the interval scale can have a specific reference point and a specific spacing. And the measurements performed in the interval scale by different measurers satisfy the ratio of the differences:

$$\frac{\tilde{y}_i - \tilde{y}_i'}{\tilde{y}_i'' - \tilde{y}_i'''} = \frac{\alpha_i y_i + c_i - (\alpha_i y_i' + c_i)}{\alpha_i y_i'' + c_i - (\alpha_i y_i''' + c_i)} = \frac{y_i - y_i'}{y_i'' - y_i'''}.$$

The above-mentioned scales (absolute scale, ratio scale, difference scale and interval scale) belong to *quantitative scales*. Naturally, the measurement results that are invariant with respect to the linear positive transformation of the general form $\tilde{y}_i = \alpha_i y_i + c_i$ inherit this property with respect to the transformations $\tilde{y}_i = a_i y_i$ or $\tilde{y}_i = y_i + c_i$. It explains why the interval scale is most "general" among the quantitative scales. In this context, all assertions established for the measurements in the interval scale remain in force for the measurements in the ratio scale and in the difference scale (and, of course, in the absolute scale).

Besides quantitative scales there exist *qualitative scales*. A typical representative of this class is the *ordinal scale* in which the measurement results are defined within a transformation $\phi_i(y_i)$ where ϕ_i denotes an arbitrary strictly increasing function. As examples, we refer to Mohs' scale for the hardness of minerals, the ordering scale for different works based on their importance, as well as various rating scales. The ordinal scales have no fixed reference point, possibly involving different spacing. Figuratively speaking, different measurers may even employ variable spacing between the marks. The statements "object 1 is α_i times greater (smaller) than object 2" and "object 1 is greater (smaller) than object 2 by the constant c_i" appear meaningless for the measurement results in the ordinal scale. Only the "greater-smaller" relation makes sense here.

All assertions established for the measurements in a qualitative scale remain in force for the measurements in a quantitative scale, but the converse fails. Thus, the quantitative scales are "richer" than the qualitative ones, as yielding more substantial assertions (though, for a narrower class of problems).

2.4.2 *Pareto Set Invariance with Respect to Strictly Increasing Transformation of Criteria*

Recall the definition of the set of Pareto optimal vectors:

$$P(Y) = \{y^* \in Y | \text{ there exists no } y \in Y \text{ such that } y \geq y^*\}.$$

The inequality $y \geq y^*$ in the definition of the Pareto set means the component-wise inequalities $y_i \geqq y_i^*$ for all $i = 1, 2, \ldots, m$, with at least one of them being strict.

Let ϕ_i be a strictly increasing numerical function of single variable defined on the whole real axis, i.e.,

$$y_i > y_i' \Leftrightarrow \phi_i(y_i) > \phi_i(y_i')$$

for all $y_i, y_i' \in R$. Obviously, the equality $y_i = y_i'$ holding for a strictly increasing function ϕ_i is equivalent to the equality $\phi_i(y_i) = \phi_i(y_i')$. Next, by the definition of this function, the inequality $y_i > y_i'$ takes place if and only if $\phi_i(y_i) > \phi_i(y_i')$ is the case.

Hence, the definition of the Pareto set does not change essentially if a strictly increasing transformation is applied to the values of criteria. In other words, the Pareto set has invariance with respect to the above transformation, and *the notion of the Pareto set can be used whenever the criteria are measured at least in the ordinal scale* (all the more, in any quantitative scale).

2.4.3 *Invariance of Theorem 2.5 with Respect to Linear Positive Transformation*

Theorem 2.5 shows how an elementary information quantum can be used for Pareto set reduction. As stated in the previous section, this reduction proceeds from the inclusions

$$C(X) \subset P_{\hat{f}}(X) \subset P_f(X), \tag{2.12}$$

where $P_{\hat{f}}(X)$ is the set of Pareto optimal alternatives in the multicriteria choice problem with the initial set of feasible alternatives X and the "new" vector criterion $\hat{f} = (\hat{f}_1, \hat{f}_2, \ldots, \hat{f}_m)$ calculated by the formulas

$$\hat{f}_j = w_j^* f_i + w_i^* f_j, \hat{f}_s = f_s \quad \text{for all } s \in I \backslash \{j\}. \tag{2.6}$$

Since the quantitative approach considered in the book presupposes measuring the criteria values in quantitative scales, the invariance of inclusions (2.12) with a linear positive transformation of the criteria is of certain practical interest. Note that, without such invariance, the suggested approach would be inapplicable to the real multicriteria problems with quantitative criteria.

Theorem 2.7 *Inclusions* (2.5) *and* (2.12) *are invariant with respect to a linear positive transformation of the criteria.*

□ First of all, observe that the inclusion $C(X) \subset \text{Ndom}\, X$ holds for any set of selectable alternatives under the hypothesis of Theorem 2.5. Moreover, the definition of the set of selectable alternatives $C(X)$ makes no mention of the criteria. Hence, this definition does not depend on the choice of the criteria scales, being invariant with respect to any transformation of the criteria.

In subsection 2.4.2, we have established the Pareto set invariance with respect to a strictly increasing transformation. A linear positive transformation is a special case of a strictly increasing transformation. Therefore, the Pareto set $P_f(X)$ from (2.12) inherits invariance with respect to a linear positive transformation of the criteria. To prove the invariance of the set $P_{\hat{f}}(X)$, it suffices to verify the invariance of the strict inequality $\hat{f}_j = w_j^* y_i + w_i^* y_j > w_j^* \bar{y}_i + w_i^* \bar{y}_j = \bar{f}_j$ incorporating the new criterion f_j, since for an arbitrary criterion f_i, $i \neq j$, the invariance of the corresponding inequalities is checked in the same elementary way as in subsection 2.4.2.

At the beginning, recall that

$$w_i^* = y_i' - y_i'', \quad w_j^* = y_j'' - y_j',$$

where $y_k' = f_k(x')$, $y_k'' = f_k(x'')$ $(k = i, j)$ and w_i^*, w_j^* are fixed positive numbers.

Replace y_k with $\tilde{y}_k = \alpha_k y_k + c_k$ $(\alpha_k > 0)$, $k = i, j$, in formula (2.6) defining the new criterion $\hat{f}_j$. This replacement yields the transformed criterion

$$\tilde{\hat{f}}_j = (\alpha_j y_j'' + c_j - \alpha_j y_j' - c_j) \cdot (\alpha_i y_i + c_i) + (\alpha_i y_i' + c_i - \alpha_i y_i'' - c_i)(\alpha_j y_j + c_j).$$

And trivial simplifications lead to

$$\tilde{\hat{f}}_j = \alpha_i \alpha_j w_j^* y_i + \alpha_i \alpha_j w_i^* y_j + C, \tag{2.15}$$

where the constant

$$C = \alpha_j w_j^* c_i + \alpha_i w_i^* c_j$$

does not depend on y_i, y_j.

Now, assume that the inequality

$$\hat{f}_j = w_j^* y_i + w_i^* y_j > w_j^* \bar{y}_i + w_i^* \bar{y}_j = \bar{f}_j \tag{2.16}$$

holds for arbitrary fixed numbers $y_i, y_j, \bar{y}_i, \bar{y}_j$. Having (2.15) in mind, multiply by the positive number $\alpha_i \alpha_j$ and add the constant C to both sides of inequality (2.16) to get

$$\tilde{\hat{f}}_j > \tilde{\bar{f}}_j = \alpha_i \alpha_j w_j^* \bar{y}_i + \alpha_i \alpha_j w_i^* \bar{y}_j + C. \tag{2.17}$$

Subsequently, inequality (2.16) implies inequality (2.17). In a similar way, inequality (2.16) can be obtained from inequality (2.17). This means the equivalence of the two inequalities. ■

Note that the degree of compromise θ_{ij} is not invariant with respect to a linear positive transformation of the criteria. Furthermore, as easily verified, the degree of compromise is not invariant with respect to the transformations $\tilde{y}_k = a_k y_k$ and $\tilde{y}_k = y_k + c_k$, $k = i,j$, which indicates of the following. *For different measurers (different DMs), the degrees of compromise may differ even if the DMs are considered in the same choice problem, have identical preferences and perform measurements in a scale of the same type*. This fact contains no contradiction, as the DMs may adopt different units of measure for the same criteria.

Really, imagine two DMs with identical preferences, who measure the values of the first criterion in USD and, RUB respectively. Suppose that the values of the second criterion are measured by them in the absolute scale (e.g., in pcs). For the DM 1 operating USD and willing to compromise 10 pcs for the gain of \$1000, the degree of compromise for the first criterion in comparison with the second one makes up

$$\theta'_{12} = \frac{10}{1000 + 10} \approx 0.01.$$

The other DM 2 operating RUB and acting in the same way must be willing to compromise 10 pcs for the gain of 60,000 RUB, since at the moment of decision-making \$1 = 60 RUB. Therefore, for DM 2 the degree of compromise constitutes

$$\theta''_{12} = \frac{10}{60000 + 10} \approx 0.00016,$$

which is considerably smaller than for DM 1. But this result is correct, since the latter operates the much more "expensive" currency than the former.

Chapter 3
Pareto Set Reduction Based on General Information Quantum

The notion of an elementary information quantum for two criteria (see Chap. 2) is extended to the general case of two groups of criteria. We study the properties of a general information quantum, demonstrating how it should be used for Pareto set reduction. To this end, one has to construct the Pareto set in terms of a new vector criterion of dimensionality that can be appreciably higher than of the original criterion.

Some geometrical illustrations are given for the choice problem with three criteria.

3.1 Definition and Properties of General Information Quantum

3.1.1 Basic Definitions

Consider two Pareto optimal vectors y', y''. By the definition of Pareto optimality, neither of the relationships $y' \geq y'', y' \leq y''$ hold. In this case, there must exist two subsets of the nonempty criteria indexes $A, B \subset I,\ A \cap B = \varnothing$, such that $y'_i > y''_i$ for all $i \in A,\ y'_j < y''_j$ for all $j \in B$, and $y'_s = y''_s$ for all $s \notin A \cup B$. If the DM prefers the first vector from this pair, i.e., $y' \succ y''$, then the second vector becomes unselected due to the exclusion axiom. Thereby, the Pareto set is reduced by the vector y''.

Definition 3.1
Let $A, B \subset I$, where $A \neq \varnothing$, $B \neq \varnothing$, and $A \cap B = \varnothing$. We say that *there is an information quantum about the DM's preference relation with the two groups of*

© Springer International Publishing AG 2018

V.D. Noghin, *Reduction of the Pareto Set*, Studies in Systems, Decision and Control 126, https://doi.org/10.1007/978-3-319-67873-3_3

criteria A and B together with the collections of positive parameters w_i^* for all $i \in A$ and w_j^* for all $j \in B$ if, for any pair of vectors $y', y'' \in R^m$ satisfying

$$\begin{array}{lll} y_i' - y_i'' = w_i^* > 0 & \text{for all} & i \in A, \\ y_j'' - y_j' = w_j^* > 0 & \text{for all} & j \in B, \\ y_s' = y_s'' & \text{for all} & s \in I \backslash (A \cup B), \end{array} \tag{3.1}$$

the relationship $y' \succ y''$ holds. In this case, the group A is more important than group B with the corresponding parameters.

In other words, given this quantum, the DM who chooses from the pair of vectors each time is willing to sacrifice the quantity w_j^* in terms of each less important criterion f_j, $j \in B$, for gaining the quantity w_i^* in terms of each more important criterion f_i, $i \in A$, the values of all other criteria being fixed.

Clearly, in the special case of $A = \{i\}$ and $B = \{j\}$, Definition 3.1 coincides with Definition 2.2 of an elementary information quantum.

Just like in the elementary case, the correlation between the quantities w_i^* and w_j^* gives a quantitative estimation for the degree of compromise of one group of criteria against the other.

Definition 3.2
Assume that there is an information quantum with the two groups of criteria A and B, and the two collections of positive parameters w_i^* for all $i \in A$ and w_j^* for all $j \in B$, respectively. The positive numbers

$$\theta_{ij} = \frac{w_j^*}{w_i^* + w_j^*} \in (0, 1), \ \text{for } i \in A \text{ and } j \in B, \tag{3.2}$$

will be called the DM's *coefficients* (or *degrees*) *of compromise* for this pair of groups of criteria.

Denote by $|A|$ and $|B|$ the numbers of elements in the sets A and B, respectively. The number of all degrees of compromise introduced by Definition 3.2 is the product $|A| \cdot |B|$. For instance, if $A = \{i\}$, then the number of the degrees coincides with the number of less important criteria $|B|$.

3.1.2 Properties of Information Quantum

The following result takes place.

Theorem 3.1 *Assume that there is a given information quantum with two collections of positive parameters w_i^* for all $i \in A$ and w_j^* for all $j \in B$. Then:*

- *There exists an information quantum with the two groups of criteria $A \cup \{k\}$ for any $k \in I \backslash (A \cup B)$ and B with the positive parameters w_i^* for all $i \in A$, w_j^* for all $j \in B$, and an arbitrary positive parameter w_k^*;*
- *There exists an information quantum with the two groups of criteria $A \cup \{k\}$ for any $k \in B$ and $B \backslash \{k\}$ with the positive parameters w_i^* for all $i \in A$, w_j^* for all $j \in B \backslash \{k\}$, and an arbitrary positive parameter w_k^*;*
- *There exists an information quantum with the two groups of criteria A and $B \backslash \{k\}$ for any $k \in B$ with the positive parameters w_i^* for all $i \in A$ and w_j^* for all $j \in B \backslash \{k\}$.*

□ According to Definition 3.1, let the relationship $y' \succ y''$ hold for vectors y', y'' satisfying (3.1).

Consider an arbitrary vector $y \in R^m$ such that

$$y_k > y'_k, \quad y_s = y'_s \text{ for all } s \in I \backslash \{k\}, k \in I \backslash (A \cup B).$$

By Axiom 3, we have the relationship $y \succ y'$. Due to the transitivity of the preference relation, this relationship in combination with $y' \succ y''$ implies $y \succ y''$. Since

$$y_i - y''_i = w_i^* > 0 \text{ for all} i \in A,$$

$$y_k > y'_k = y''_k,$$

$$y''_j - y'_j = w_j^* > 0 \text{ for all} j \in B,$$

$$y_s = y''_s \text{ for all } s \in I \backslash (A \cup B \cup \{k\}),$$

and the difference $w_k^* = y_k - y''_k$ can be any positive number, the proof of the first statement is complete.

To verify the second statement of the theorem, introduce the vector y with the components

$$y_k > y''_k, \; y_s = y''_s \text{ for all} s \in I \backslash \{k\}, k \in B.$$

As above, this vector satisfies the relationship $y' \succ y''$, which establishes the second statement.

And finally, the reader can show that the third statement is true without much effort, following the same line of reasoning as before. ■

According to this result, the *group B of less important criteria can be reduced, while the group A of more important criteria can be extended.* And the new parameters corresponding to the added criteria can take any positive values.

As follows from the general considerations, if the DM is willing to lose something in terms of the less important criteria for gaining in terms of the more important criteria, then the DM obviously agrees with a smaller loss and with a greater gain in terms of these criteria, too. And the following result takes place.

Theorem 3.2 *Assume that there is a given information quantum with the two groups of criteria* $A = \{i_1, i_2, \ldots, i_k\}$ *and* $B = \{j_1, j_2, \ldots, j_l\}$ *together with two given collections of positive parameters* $w^*_{i_s}$ *for all* $s = 1, 2, \ldots, k$ *and* $w^*_{j_s}$ *for all* $s = 1, 2, \ldots, l$*. Then there is a given information quantum with the same two groups of criteria A and B together with any pair of the collections of positive parameters* w'_{i_s} *for all* $s = 1, 2, \ldots, k$ *and* w'_{j_s} *for all* $s = 1, 2, \ldots, l$ *satisfying the inequality*

$$(w'_{i_1}, \ldots, w'_{i_k}, -w'_{j_1}, \ldots, -w'_{j_l}) \geq (w^*_{i_1}, \ldots, w^*_{i_k}, -w^*_{j_1}, \ldots, -w^*_{j_l}).$$

In other words, if the group of criteria A is more important than the group of criteria B with the degree of compromise θ_{ij} *for all* $i \in A$ *and all* $j \in B$*, then the former is more important than the latter with the degree of compromise* $\theta'_{ij} < \theta_{ij}$ *for all* $i \in A$ *and all* $j \in B$*.*

The proof of Theorem 3.2 resembles that of Theorem 2.1, thus being omitted here.

In addition, it is possible to define the relation of the incomparably higher importance of one group of criteria against another group. Notably, if any positive number $\theta_{ij} \in (0, 1)$ (for all $i \in A$ and $j \in B$) is the degree of compromise for the group of criteria A against the group of criteria B, we say that *the first group of criteria is incomparably more important than the second group*.

In Chap. 2, we have obtained a characterization of the lexicographical relations in terms of a sequence of incomparably more important criteria (see Theorem 2.2 for details). A new characterization in terms of groups of criteria is given below.

Theorem 3.3 *The binary relation* $\succ$ *defined on space* R^m *that satisfies Axioms 2 and 3 is a lexicographic relation if and only if criterion 1 is incomparably more important than the group of the successive criteria* $\{2, 3, \ldots, m\}$*, criterion 2 is incomparably more important than the group of the successive criteria* $\{3, \ldots, m\}$ *, …, and criterion* $(m - 1)$ *is incomparably more important than criterion m.*

□ Necessity. Let the relation $\succ$ be lexicographic. By the definition of a lexicographical relation, for arbitrary vectors $y', y'' \in R^m$ we have the logical propositions

(1) $y'_1 > y''_1 \quad \Rightarrow \quad y' \succ y''$,
(2) $y'_1 = y''_1, \; y'_2 > y''_2 \quad \Rightarrow \quad y' \succ y''$,
(3) $y'_1 = y''_1, \; y'_2 = y''_2, \; y'_3 > y''_3 \quad \Rightarrow \quad y' \succ y''$,
……………………………………………………
(m) $y'_i = y''_i, \; i = 1, 2, \ldots, m - 1; \quad y'_m > y''_m \quad \Rightarrow \quad y' \succ y''$.

Proposition (1) implies that criterion 1 is incomparably more important than the group of the other criteria. Really, according to proposition (1), for arbitrary vectors $y', y'' \in R^m$ with the positive differences $y''_2 - y'_2, \ldots, y''_m - y'_m$, all numbers

$$\theta_{12} = \frac{y''_2 - y'_2}{y'_1 - y''_1 + y''_2 - y'_2},$$
$$\theta_{13} = \frac{y''_3 - y'_3}{y'_1 - y''_1 + y''_3 - y'_3},$$
$$\ldots\ldots\ldots\ldots\ldots\ldots\ldots\ldots\ldots\ldots$$
$$\theta_{1m} = \frac{y''_m - y'_m}{y'_1 - y''_1 + y''_m - y'_m},$$

are the degrees of compromise. Moreover, as the above-mentioned differences together with $y'_1 - y''_1$ can take all possible values, the degrees of compromise represent arbitrary numbers completely filling the interval $(0, 1)$. As a result, criterion 1 appears incomparably more important than the group of the other criteria.

Similarly, using proposition (2), we conclude that criterion 2 is incomparably more important than the group of the successive criteria $\{3, \ldots, m\}$, and so on; in the final analysis, the incomparably higher importance of criterion $(m-1)$ against criterion m follows from proposition $(m-1)$.

Sufficiency. Let criterion 1 be incomparably more important than the group of all other criteria $\{2, 3, \ldots, m\}$, let criterion 2 be incomparably more important than the group of the successive criteria $\{3, \ldots, m\}$, and so on. Choose two arbitrary vectors $y', y'' \in R^m$ satisfying the inequality $y'_1 > y''_1$. To prove proposition (1), it is necessary to check that the relationship $y' \succ y''$ holds.

If, in addition to the inequality $y'_1 > y''_1$, we have $y'_i \geqq y''_i$, $i = 2\ldots, m$, then the relationship $y' \succ y''$ holds by the Pareto axiom.

Consider the case where, together with the inequality $y'_1 > y''_1$, the inequality $y'_s < y''_s$ takes place for some one or several $s \in \{2, \ldots, m\}$. Introduce a vector y such that $y_1 = y'_1$, $y_s = y'_s - 1$ for all s above and $y_k = y''_k - 1$ for all other indexes k. Obviously, the inequality $y' \geq y$ is satisfied. Hence, according to the Pareto axiom, we obtain the relationship $y' \succ y$. Only the first component of the vector y is greater than that of the vector y'', and all other components of the former are smaller than their counterparts in the vector y''. It follows that $y \succ y''$, since criterion 1 is incomparably more important than the collection of all other criteria. Due to the transitivity of the relation $\succ$, the relationships $y' \succ y$ and $y \succ y''$ yield the desired result $y' \succ y''$.

In a similar manner, using the Pareto axiom and the incomparably higher importance of criterion 2 against the group of the successive criteria $\{3, \ldots, m\}$, we verify proposition (2), and so on.

Following this procedure, in the final analysis we establish the truth of proposition $(m-1)$. Proposition (m) follows from Axiom 3. ■

3.2 Pareto Set Reduction Using Information Quantum

3.2.1 Simplified Definition of Information Quantum

For verifying that there is given information quantum, by Definition 3.1 the DM has to compare infinitely many pairs of vectors y', $y'' \in R^m$ satisfying relationships (3.1) with some positive parameters w_i^*, w_j^*. Obviously, such verification appears non-implementable in practice. Actually, just like in the case of two criteria (see Theorem 2.4), this verification is not required, since the preference relation possesses invariance. It suffices to check relationships (3.1) merely for a fixed pair of vectors y', y'', as indicated by the following result.

Theorem 3.4 *Owing to the invariance of the preference relation* $\succ$, *the vectors* y', y'' *in Definition* 3.1 *can be assumed fixed. Particularly,*

$$\begin{aligned} y_i' &= w_i^* \quad \text{for all } i \in A, \\ y_j' &= -w_j^* \quad \text{for all } j \in B, \\ y_s' &= 0 \quad \text{for all } s \in I \backslash \{A \cup B\}, \end{aligned} \tag{3.3}$$

and $y'' = 0_m$.

The proof involves the same scheme as in Theorem 2.4.

Since the preference relation $\succ$ is invariant with respect to a linear positive transformation, we use Theorem 3.4 for introducing a simplified definition of an information quantum. Actually, it is equivalent to Definition 3.1.

Definition 3.3 Let $A, B \subset I$, where $A \neq \varnothing$, $B \neq \varnothing$, and $A \cap B = \varnothing$. We say that there is *an information quantum about the DM's preference relation with two given groups of criteria A and B together with two collections of positive parameters* w_i^* *for all* $i \in A$ *and* w_j^* *for all* $j \in B$ if a vector y' of form (3.3) satisfies the relationship $y' \succ 0_m$.

For instance, if the vector $(0.7, -0.3, 1)$ is preferred to the vector $(0, 0, 0)$, then the group composed of criteria 1 and 3 is more important than the group composed of criterion 2 only, and the corresponding degrees of compromise are

$$\theta_{12} = \frac{0.3}{0.7 + 0.3} = 0.3, \quad \theta_{32} = \frac{0.3}{1 + 0.3} \approx 0.23.$$

3.2.2 Pareto Set Reduction Based on Information Quantum

An information quantum being available, one can apply the next theorem to eliminate from the Pareto set the vectors that are not selectable for sure.

Let us introduce the following convenient term. Namely, let $a^i \in R^m$, $\lambda_i \in R$, $i = 1, 2, \ldots, n$. A linear combination $\sum_{i=1}^{n} \lambda_i a^i$ will be called the *N-combination* if $(\lambda_1, \lambda_2, \ldots, \lambda_n) \geq 0_n$.

Theorem 3.5. *Assume that $A, B \subset I$, where $A \neq \varnothing$, $B \neq \varnothing$, $A \cap B = \varnothing$, and there is a given information quantum about the DM's preference relation with two given groups of criteria A and B together with the collections of positive parameters w_i^* for all $i \in A$ and w_j^* for all $j \in B$. Then for any set of selectable vectors $C(Y)$ we have*

$$C(Y) \subset \hat{P}(Y) \subset P(Y), \tag{3.4}$$

where $P(Y)$ is the set of Pareto optimal vectors in the multicriteria problem with the initial set of feasible alternatives X and the initial vector criterion f, while $\hat{P}(Y)$ is the set of feasible vectors corresponding to the set of Pareto optimal alternatives in the multicriteria problem with the set X and the new p-dimensional vector criterion g, $p = m - |B| + |A| \cdot |B|$ (i.e., $\hat{P}(Y) = f(P_g(X))$) composed of all components f_i of the vector criterion f that satisfy $i \in I \backslash B$ and of the components

$$g_{ij} = w_j^* f_i + w_i^* f_j \text{ for all } i \in A \text{ and all } j \in B, \tag{3.5}$$

or

$$g'_{ij} = \theta_{ij} f_i + (1 - \theta_{ij}) f_j \quad \text{for all } i \in A \text{ and all } j \in B. \tag{3.5'}$$

□ I. Denote by K the acute convex cone of the cone preference relation $\succ$. By hypothesis, the vector y' described by equalities (3.3) satisfies the relationship $y' \succ 0_m$. The latter is equivalent to the inclusion $y' \in K$. According to Theorem 2.1, we have the inclusion $R_+^m \subset K$.

Introduce the set M as the collection of all N-combinations of a collection of vectors $e^1, e^2, \ldots, e^m, y'$, where $e^1, e^2, \ldots, e^m$ are the unit vectors of space R^m. The set M is a convex cone not containing the origin, since the coefficients of the linear combinations cannot be zero simultaneously.

Moreover, M is an acute cone. If we suppose the opposite, then there exist two N-combinations $y = \sum_{i=1}^{m} \lambda_i e^i + \lambda_{m+1} y'$ and $-y = \sum_{i=1}^{m} \lambda'_i e^i + \lambda'_{m+1} y'$, such that their sum

$$y - y = \sum_{i=1}^{m} (\lambda_i + \lambda'_i) e^i + (\lambda_{m+1} + \lambda'_{m+1}) y' = 0_m$$

is the N-combination too. If $\lambda_{m+1} + \lambda'_{m+1} = 0$, then the last system of equation gives $\lambda_i + \lambda'_i = 0$ for all $i = 1, 2, \ldots, m$. That is impossible, since $y - y$ is the N-combination. If $\lambda_{m+1} + \lambda'_{m+1} > 0$, then $\lambda_i + \lambda'_i < 0$ for all $i \in A$ that is impossible too.

Introduce the *dual*[1] *cone* (without the origin) for the cone M:

$$C = \{y \in R^m |\ \langle z, y \rangle \geqq 0 \text{ for all } z \in M\} \backslash \{0_m\}.$$

According to duality theory of convex analysis (see [62]), the generators of the cone C are the inner normals to the $(m-1)$-dimensional faces of the cone M. And conversely, the generators of the cone M are the inner normals to the $(m-1)$-dimensional faces of the cone C.

II. Two cases are possible, namely, $|A| > 1$ and $|A| = 1$. In the first case, the generators of the cone M are all vectors $e^1, e^2, \ldots, e^m, y'$, since none of these vectors can be expressed as the N-combination of the other vectors from this collection. In the second case (when $A = \{i\}$), the vector e^i can be expressed as the N-combination of the vector y' and all vectors e^s with $s \in B$. Hence, here the generators of the cone M are the vectors $e^1, e^2, \ldots, e^m, y'$ except the vector e^i. Next, we analyze these cases by turn.

Since the generators of the cone M are the vectors $e^1, e^2, \ldots, e^m, y'$, then the set of nonzero solutions to the system of linear homogeneous inequalities

$$\langle e^i, y \rangle \geqq 0 \text{ for all } i \in I,$$

$$\langle y', y \rangle \geqq 0, \tag{3.6}$$

coincides with the dual cone C.

Find the fundamental system of solutions to the system of linear inequalities (3.6). It must be a system of vectors whose set of linear nonnegative combinations matches the solution set of the system of inequalities (3.6). And none vector of the fundamental system can be represented as the N-combination of the other vectors in this system.

At the beginning, specify some collection of solutions to the system of linear inequalities (3.6). First of all, note that for $i \in I \backslash B$ each unit vector e^i of space R^m solves (3.6). Next, introduce the vectors

$$e^{ij} = w_j^* e^i + w_i^* e^j \quad \text{for all } i \in A \text{ and all } j \in B.$$

The components of these vectors are nonnegative, and therefore they all satisfy the inequality $\langle e^i, y \rangle \geqq 0$ for each $i \in I$. Moreover, they satisfy the last inequality $\langle y', y \rangle \geqq 0$ in (3.6), since

---

[1]Dual cones are also considered in Sect. 4.3.

$$\langle y', e^{ij} \rangle = y'_i w_j^* + y'_j w_i^* = 0 \quad \text{for all } i \in A \text{ and } j \in B.$$

Consequently, the collection composed the vectors e^i for all $i \in I \backslash B$ and the vectors e^{ij} for all $i \in A$ and $j \in B$ actually belongs to the dual cone C. In addition, as easily checked, none of the vectors from this collection can be represented as the N-combination of the other vectors. The total number of all vectors in this collection is $p = m - |B| + |A| \cdot |B|$.

For verifying that the above collection of vectors forms the fundamental system of solutions to the system of linear inequalities (3.6), it remains to make sure that (3.6) do not have other solutions except all nonnegative linear combinations of the vectors of this collection.

To this end, together with (3.6), consider the corresponding system of $(m + 1)$ linear equations

$$\begin{aligned} \langle e^i, y \rangle &= 0 \text{ for all } i \in I, \\ \langle y', y \rangle &= 0. \end{aligned} \tag{3.7}$$

The desired fundamental system of solutions to the system of linear inequalities (3.6) is contained among the one-dimensional solutions to the subsystems of the linear equations (3.7).

In the collection of the vectors $e^1, e^2, \ldots, e^m, y'$ answering (3.7), we are concerned with the subcollections of rank $m - 1$. Exactly the subsystems answering these subcollections have one-dimensional solutions. Among the obtained one-dimensional solutions, choose the ones satisfying the system of inequalities (3.6). The resulting vectors form the required fundamental system of solutions to the system of inequalities (3.6).

Since elimination of any pair of vectors from the collection $e^1, e^2, \ldots, e^m, y'$ leads to a subsystem of rank $m - 1$, let us successively remove two equations from (3.7).

If the last equation of (3.7) is among the removed ones, for obtaining the one-dimensional solutions one has to eliminate another equation of the form $\langle e^i, y \rangle = 0$. The resulting subsystems possess the unit vectors $e^1, e^2, \ldots, e^m$ as their solutions within a positive factor. Clearly, in this collection, the only vectors that satisfy the system of inequalities (3.6) are the ones with indexes not belonging to B.

Let the subsystem includes the last equation of (3.7). If any two equations of the form $\langle e^i, y \rangle = 0$ for $i \in A$ and $\langle e^j, y \rangle = 0$ for $j \in B$ are removed, then the resulting "truncated" subsystems among their nonzero solutions have only the one-dimensional solutions. Among them, choose the vectors e^{ij} for all $i \in A$ and all $j \in B$. All these vectors satisfy the system of inequalities (3.6), as shown earlier. By eliminating the pairs of equations $\langle e^i, y \rangle = 0$ with index i belonging to the set A or B only, we will not construct the subsystems having nonzero solutions. If the removed equations are only of the form $\langle e^i, y \rangle = 0$ for $i \in I \backslash (A \cup B)$, then we obtain no additional nonzero one-dimensional solutions.

This means that the collection of vectors compiled from e^i for all $i \in I \backslash B$ and e^{ij} for all $i \in A$ and all $j \in B$ forms the fundamental system of solutions to the system of linear inequalities (3.6) and any solution to (3.6) can be represented as the nonnegative linear combination of the vectors from this collection. In the sequel, this collection will be denoted by $a^1, a^2, \ldots, a^p$. Consideration of the first case is complete.

Let us treat the second case in a few words. For $A = \{i\}$, the line of reasoning is same yet somewhat simpler than before. In this case, we have to consider a system of m equations that differs from (3.7) in the absence of the equation $\langle e^i, y \rangle = 0$ corresponding to the unit vector e^i. This fact explains why it is necessary to eliminate only one equation from (3.7) to get the same fundamental system of solutions to the system of linear inequalities (3.6) as in the first case.

III. According to the above result, the solution set to the system of linear inequalities (3.6), i.e., the cone C coincides with the set of all N-combinations of the vectors $a^1, a^2, \ldots, a^p$. Therefore, for the vector z the inclusion $z \in C$ takes place if and only if this vector can be represented as some N-combination of the vectors from the above-mentioned collection.

Owing to the last circumstance, for an arbitrary fixed vector $y \neq 0_m$, the inequalities

$$\langle z, y \rangle \geqq 0 \text{ for all } z \in C \tag{3.8}$$

appear equivalent to the inequalities

$$\langle a^i, y \rangle \geq 0, \quad i = 1, 2, \ldots, p, \tag{3.8$'$}$$

where the sign $\geq$ indicates strict inequality at least for one index $i \in \{1, 2, \ldots, p\}$. Really, each vector $z \in C$ can be expressed as some N-combination of the vectors $a^1, a^2, \ldots, a^p$, e.g., $z = \lambda_1 a^1 + \lambda_2 a^2 + \cdots + \lambda_p a^p$. And if the vector y satisfies inequalities (3.8′), then through multiplying these inequalities by the nonnegative numbers $\lambda_1, \lambda_2, \ldots, \lambda_p$ and performing the termwise addition of the resulting inequalities, we get

$$\left\langle \sum_{i=1}^{p} \lambda_i a^i, y \right\rangle = \langle z, y \rangle \geqq 0$$

from (3.8). Conversely, inequality (3.8) implies inequality (3.8′), since $a^i \in C$ for all $i = 1, 2, \ldots, p$.

Inequalities (3.8′) cannot hold as equalities all simultaneously. Indeed, if inequalities (3.8′) hold as equalities for the nonzero vector y, then these equalities also take place for the opposite vector $-y$. Hence, the cone dual to C is not acute. But this dual cone has the form

$$M = \{y \in R^m | \langle z, y \rangle \geqq 0 \text{ for all } z \in C\} \backslash \{0_m\}, \tag{3.9}$$

since C is dual to the cone M.[2] Thereby, we have arrived at a contradiction: the cone M is not acute. This means that, for the nonzero vector y, inequalities (3.8′) cannot hold as equalities all simultaneously.

Based on the established equivalence of inequalities (3.8) and (3.8′), from (3.9) we deduce that

$$y \in M \Leftrightarrow \langle a^i, y \rangle \geq 0, \quad i = 1, 2, \ldots, p \tag{3.10}$$

IV. Let us proceed to the final stage of the proof. The inclusions

$$R^m_+ \subset M \subset K$$

imply

$$\text{Ndom}\, Y \subset \hat{P}(Y) \subset P(Y), \tag{3.11}$$

where

$$\hat{P}(Y) = \{y^* \in Y | \text{ there exists no } y \in Y \text{ such that } y - y^* \in M\}$$

is the set of nondominated elements of the set Y ordered by the cone relation with the acute convex cone M.

Choose arbitrarily $y = f(x)$, $y^* = f(x^*)$ such that $f(x) \neq f(x^*)$ for some $x, x^* \in X$. Owing to equivalence (3.10), the inclusion $f(x) - f(x^*) \in M$ is true if and only if

$$\langle a^i, f(x) - f(x^*) \rangle \geq 0, \quad i = 1, 2, \ldots, p,$$

or, after trivial transformations,

$$\langle a^i, f(x) \rangle \geq \langle a^i, f(x^*) \rangle, \quad i = 1, 2, \ldots, p.$$

Using the specific form of the vectors $a^1, a^2, \ldots, a^p$, rewrite the last inequalities as

$$g(x) \geq g(x^*),$$

where g is the p-dimensional vector function mentioned in the statement of Theorem 3.5. Subsequently, $\hat{P}(Y) = f(P_g(X))$, i.e., $\hat{P}(Y)$ forms the set of all vectors corresponding the set of Pareto optimal alternatives in the multicriteria choice problem with the initial set of feasible alternatives X and the vector criterion g.

---

[2]If M represents a polyhedral cone, then the cone M is dual to the dual cone C, see [57].

To finish the proof of inclusions (3.4), it remains for any $C(Y)$ to add the inclusion $C(Y) \subset \mathrm{Ndom}\, Y$ to (3.11), which actually holds by Lemma 1.2.

And finally, in formula (3.4) the functions g_{ij} of form (3.5) can be replaced by g'_{ij} of form (3.5′): the latter are obtained from the former using the division by the positive constant $w_i^* + w_j^*$. ■

One may easily reformulate this result in terms of alternatives. Notably, the following theorem is valid.

Theorem 3.5 (in terms of alternatives). *Assume that there is a given information quantum about the DM's preference relation with the two groups of criteria A and B, and the collections of positive parameters w_i^* for all $i \in A$ and w_j^* for all $j \in B$. Then for any set of selectable alternatives $C(X)$ we have*

$$C(X) \subset P_g(X) \subset P_f(X), \tag{3.12}$$

where $P_f(X)$ is the set of Pareto optimal alternatives in the multicriteria problem with the set of feasible alternatives X and the vector criterion f, while $P_g(X)$ is the set of Pareto optimal alternatives in the problem with the set X and the new p-dimensional vector criterion g stated in the previous theorem.

According to the established result, the new vector criterion g consists of $p = m - |B| + |A| \cdot |B| \geqq m$ components. Hence, the number of new criteria may coincide with the number m of "old" criteria or exceed it.

Corollary 3.1 *Under the hypothesis of Theorem 3.5, the equality $p = m$ holds if and only if $|A| = 1$.*

□ Let $p = m - |B| + |A| \cdot |B| = m$. Then $|A| \cdot |B| = |B|$, and therefore $|A| = 1$. Conversely, if $|A| = 1$, then $p = m - |B| + 1 \cdot |B| = m$. ■

Example 3.1 Consider the multicriteria choice problem with ten criteria ($m = 10$), where a certain half of the criteria is more important than the other, i.e., $|A| = |B| = 5$. In this case, by Theorem 3.5 we have $p = 10 - 5 + 5 \cdot 5 = 30$. And so, the new vector criterion g contains the five old criteria and the twenty five new ones calculated by formula (3.5).

The next result shows under which conditions the number of components in the new vector criterion is maximum possible.

Corollary 3.2 *Under hypotheses of Theorem 3.5, the maximum value of p is reached if*

$$|A| = \left[\frac{m+1}{2}\right], \quad |B| = m - |A|,$$

where $[\,\cdot\,]$ denotes the integer part operator.

□ Let $x = |A|$, $y = |B|$. Consider the maximization problem

$$p = m - y + xy \to \max$$

subject to the condition $x + y \leqq m$. Obviously, the maximum in this optimization problem is reached only under the equality $x + y = m$. Using the latter, express y through x and substitute the result into p to obtain $p = m - (m - x) + x(m - x) = x(m + 1 - x)$. This quadratic function of one variable x takes the maximum value at the point $x = \frac{m+1}{2}$. If m is an odd number, then x becomes an integer. For an even number m, the integer maximum is reached at the closest integer $|A| = \left[\frac{m+1}{2}\right]$ (just like at $|A| = \left[\frac{m+2}{2}\right]$). ■

Corollary 3.2 demonstrates that, in Example 3.1 with $m = 10$, the maximum possible number of components in the new vector criterion makes up 30, being reached if a certain half of criteria is more important than the other or if a certain group of six criteria is more important than the residual group of four criteria.

Theorem 2.7 has established the invariance of inclusions (2.12) and (2.15) with respect to a linear positive transformation of criteria in the case of an elementary information quantum. Interestingly, the formulas used to define the degrees of compromise and recalculate the new criteria are identical for two criteria and two groups of criteria. And so, the arguments adopted in the proof of Theorem 2.7 can be involved in the case of two groups of criteria. This gives the following result.

Theorem 3.6 *Inclusions* (3.4) *and* (3.12) *are invariant with respect to a linear positive transformation of the criteria* $f_1, f_2, \ldots, f_m$.

Hence, Theorem 3.5 can be applied to all multicriteria choice problems where the above criteria are measured in quantitative scales (i.e., interval scales, ratio scales and difference scales).

3.3 Geometrical Illustrations to the Problem with Three Criteria

3.3.1 *Tricriteria Problem in General Form*

An information quantum about the DM's preference relation in the bicriteria problem may only have the form where the groups of criteria represent singletons. In this case, the number of new criteria (the parameter p) coincides with the number of "old" criteria, i.e., $p = 2$. In other words, in the bicriteria problem the consideration of an information quantum using Theorem 3.5 does not increase the number of criteria (as a matter of fact, the same conclusion follows from the results of Chap. 2).

Consider the problem with $m = 3$. Assume that there is an information quantum in which the group A consists of f_1, f_2 and the group B of the criterion f_3. According to Definition 3.3, this means that the inclusion $y' \succ 0_3$ holds for some vector $y' = (w_1^*, w_2^*, -w_3^*) = OD$ under certain positive parameters w_1^*, w_2^*, w_3^*

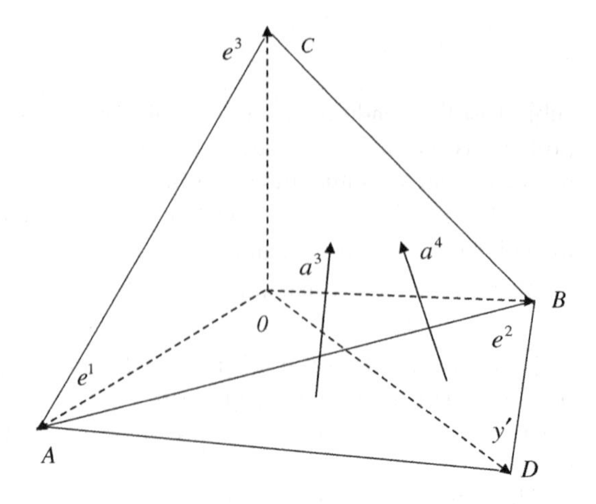

Fig. 3.1 Geometry of the tricriteria problem

(see Fig. 3.1). The specific values of these parameters make no sense for further exposition.

The nonnegative octant R^3_+ is the acute convex cone $OABC$ (without the origin) generated by the unit vectors $e^1 = OA$, $e^2 = OB$ and $e^3 = OC$. This cone has three two-dimensional facets representing the corresponding parts of the coordinate planes OBC, OAC and OAB. The convex cone M generated by the unit vectors of space R^3 and the vector y' is an acute convex cone (without the origin) having four two-dimensional facets, namely, OBC, OAC, OAD and OBD. The normal vectors (the inner normal of the cone M), namely, the vectors a^1, a^2, a^3, a^4, represent the generators of the cone C that is dual to M. Here

$$a^1 = e^1 \perp OBC, \quad a^2 = e^2 \perp OAC, \quad a^3 \perp OAD, \quad a^4 \perp OBD.$$

Since the three-dimensional cone M has four two-dimensional facets, then the dual cone C is generated by the four vectors e^1, e^2, a^3, a^4. Hence, in this case, the new vector criterion g contains four components. Really, it follows from Theorem 3.5 that $p = 3 - 1 + 2 \cdot 1 = 4$.

In the current example, the initial number of criteria $m = 3$ has been increased by 1 after taking into account the available information quantum.

Now, consider another case. Let one of the criteria be more important than the group of the other two criteria. As easily calculated, $p = 3 - 2 + 1 \cdot 2 = 3$, i.e., the number of the new criteria coincides with the number of the "old" ones. The same situation occurs if one of the criteria is more important than the other.

In fact, we have exhausted all possible separations of the criteria in terms of their importance, and it is possible to draw the following conclusion. *In the tricriteria problem, taking into account an information quantum based on Theorem* 3.5 *may increase the number of criteria only by 1 and only if the group of two criteria is more important that the third criterion.*

3.3.2 Case of Linear Criteria

Consider the tricriteria problem where the set of feasible alternatives is a subset of the vector space R^3 (viz., $X \subset R^3$) and all criteria are linear

$$f_1(x) = \langle c^1, x \rangle, \quad f_2(x) = \langle c^2, x \rangle, \quad f_3(x) = \langle c^3, x \rangle$$

where $c^1, c^2, c^3, x \in R^3$. The cone generated by the vectors c^1, c^2, c^3 (the gradients of the linear functions f_1, f_2, f_3) is called the *cone of goals*. Let these vectors be noncomplanar and have the form illustrated by Fig. 3.2. They generate a certain three-dimensional three-facet cone.

Assume that criterion 1 is more important than the group composed of criteria 2 and 3 with the degrees of compromise $\theta_{12} = \theta_{13} = 0.5$. In this case, by Theorem 3.5, taking into account such information requires considering a new multicriteria problem with criterion 1 of the same form and the new criteria of the form $g_{12}(x) = \langle c^2_{new}, x \rangle$ and $g_{13}(x) = \langle c^3_{new}, x \rangle$ instead of the original less important criteria 2 and 3 (see Fig. 3.3). Therefore, the cone of goals generated by the gradients of the goal functions in the new multicriteria problem has three edges and three facets (just like in the original multicriteria problem), but appears substantially narrower than the initial cone generated by the vectors c^1, c^2 and c^3.

And now, suppose that the group formed by criteria 2 and 3 is more important than criterion 1, and the degrees of compromise make up $\theta_{21} = \theta_{31} = 0.5$. According to Theorem 3.5, to take into account this information quantum, one has to consider a new multicriteria problem with criteria 2 and 3 of the same form and

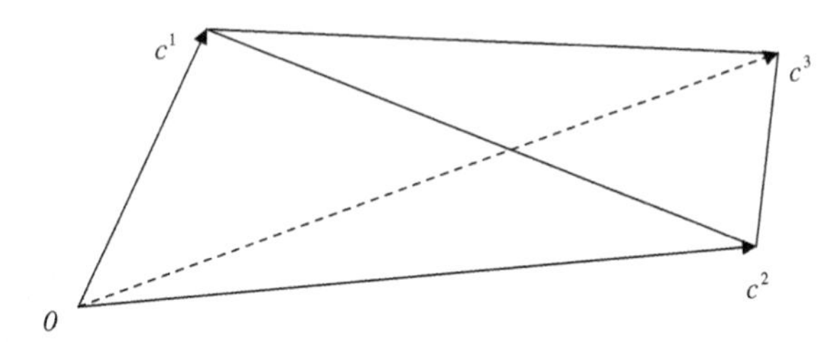

Fig. 3.2 Cone of goals

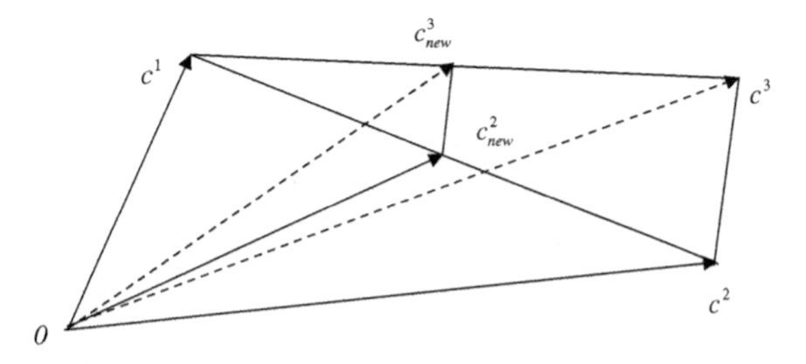

Fig. 3.3 Vectors $c^1, c^2, c^3, c^1_{new}, c^2_{new}$.

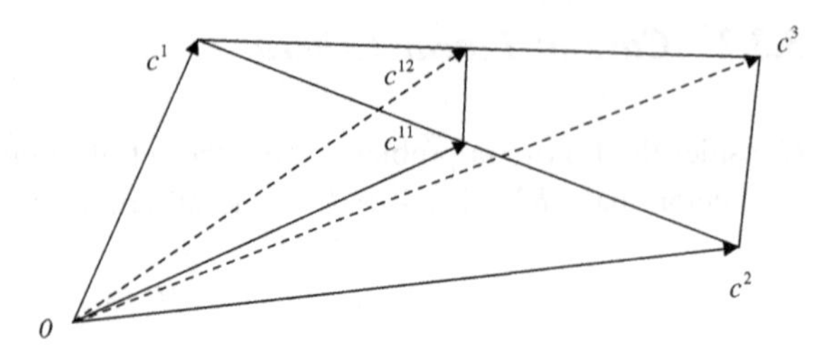

Fig. 3.4 Vectors $c^1, c^2, c^3, c^{11}, c^{12}$.

the new criteria of the form $g_{21} = \langle c^{11}, x \rangle$ и $g_{31} = \langle c^{12}, x \rangle$ instead of the original criterion 1, see Fig. 3.4.

Obviously, the cone of goals formed by the gradients c^{11}, c^{12}, c^3, c^4 of the components of the new vector criterion has four generators and four facets.

Chapter 4
Pareto Set Reduction Using Elementary Collections of Information Quanta

This chapter focuses on the application of some "simple" collections of information quanta. We establish that some collections can be inconsistent. Therefore, first the issue of consistency for an elementary collection of information quanta is considered, and then the corresponding definition is introduced and several consistency criteria in different forms are obtained.

In addition, we suggest the notion of mutually dependent and mutually independent information quanta, as well as formulate a series of results on Pareto set reduction using some consistent collections of information quanta.

4.1 Consistent Collections of Information Quanta

4.1.1 Preliminary Analysis

Let $A, B \subset I$, where $A \neq \varnothing, B \neq \varnothing, A \cap B = \varnothing$. According to Definition 3.3, by specifying a pair of vectors $y', y'' \in R^m$ with the components

$$y'_i - y''_i = w^*_i \quad \text{for all} \;\; i \in A,$$

$$y''_j - y'_j = w^*_j \quad \text{for all} \;\; j \in B,$$

$$y'_s = y''_s \quad \text{for all} \;\; s \in I \backslash (A \cup B),$$

that satisfy the relationship $y' \succ y''$, we mean that the group of criteria A is more important than the group of criteria B with the two collections of positive parameters w^*_i for all $i \in A$ and w^*_j for all $j \in B$. Since $A \neq \varnothing$ and $B \neq \varnothing$, the vector $y' - y''$ has at least one positive and one negative components. Introduce the set of all such vectors, denoting it by

© Springer International Publishing AG 2018

V.D. Noghin, *Reduction of the Pareto Set*, Studies in Systems, Decision and Control 126, https://doi.org/10.1007/978-3-319-67873-3_4

$$N^m = R^m \backslash \left[R^m_+ \cup (-R^m_+) \cup \{0_m\} \right].$$

Assume that we have identified a pair of different vectors $u, v \in R^m$ such that, for the DM, the vector u is preferable to the vector v: $u \succ_Y v$. By the transitivity axiom, the last relationship is equivalent to $u \succ v$. Let $u - v \in N^m$. Designate by A and B the index sets of the positive and negative, respectively, components of the vector $u - v$. Obviously, $A \neq \varnothing, B \neq \varnothing, A \cap B = \varnothing$. Hence, a given arbitrary pair of vectors $u, v \in R^m$ satisfying the relationships $u \succ v$ and $u - v \in N^m$ can be treated as information that the group of criteria A is more important than the group of criteria B with the two corresponding collections of positive parameters. Thereby, any pair of vectors $u, v \in R^m$ satisfying the relationship $u - v \in N^m$ may give a certain information quantum about the DM's preference relation under a specific condition (viz., the relationship $u \succ v$).

Now, assume that there is a finite collection of such pairs of vectors:

$$u^i, v^i \in R^m, \quad u^i - v^i \in N^m, \quad i = 1, 2, \ldots, k. \tag{4.1}$$

Then the following question seems natural. May these pairs of vectors specify a certain collection of information quanta? Simple examples show that, in the general case, the answer is negative.

Example 4.1 Let $m = 2, k = 2$ and

$$u^1 = (1, -3), u^2 = (-2, 1), v^1 = v^2 = 0_2.$$

Suppose that this collection of two pairs of vectors defines two elementary information quanta so that the relationships $u^1 \succ v^1$ and $u^2 \succ v^2$ hold. The termwise addition of these relationships yields $u^1 + u^2 \succ v^1 + v^2$ or, equivalently, $(-1, -2) \succ 0_2$. On the other hand, we have the relationship $0_2 \succ (-1, -2)$, since $0_2 \geq (-1, -2)$. The two relationships $(-1, -2) \succ 0_2$ and $0_2 \succ (-1, -2)$ are inconsistent with the asymmetrical property of the relation $\succ$. And so, for the above pair of vectors the relationships $u^1 \succ v^1$ and $u^2 \succ v^2$ do not hold simultaneously for any binary relation satisfying Axioms 2–4.

4.1.2 *Definition of Consistent Collection of Vectors*

Definition 4.1 Consider a given collection of the pairs of vectors (4.1). We say that this collection is *consistent* if there exists at least one binary relation $\succ$ obeying Axioms 2–4 such that the relationships $u^s \succ v^s, s = 1, 2, \ldots, k$, hold. Whenever the consistent pairs of vectors define a corresponding collection of information quanta, this information will be called *consistent*.

The consistency of the pairs of vectors (4.1) is a necessary condition under which there exists a given collection of information quanta at least in one multicriteria choice problem (at least for one DM).

While solving real multicriteria choice problems arising in applications, one disposes of a whole family of different quanta and the vectors specifying them may form an inconsistent collection. This is due to the fact that the existing information about the DM's preference relation generally contains uncertainty, merely reflecting his desired (not actual) preference pattern. Moreover, the DM himself (without wishing!) sometimes slightly deviates from the class of multicriteria choice problems satisfying Axioms 2–4; in this case, the DM's behavior should be corrected by announcing the inconsistency of his preferences expressed via the collection of information quanta.

In any event, if the choice process based on information about the DM's preference relation involves a collection of information quanta, this collection must be verified for consistency. Such a verification procedure requires appropriate tools, since Definition 4.1 is of no assistance here.

4.1.3 Criteria of Consistency

In this subsection, we give three criteria of consistency for a finite collection of vectors specifying a collection of information quanta. One of them has geometrical form, another is stated in algebraic terms, and the third criterion represents an assertion facilitating further program coding.

Theorem 4.1 (geometrical criterion of consistency) . *For a collection of the pairs of vectors* (4.1) *to be consistent, a necessary and sufficient condition is that the cone generated by the vectors*

$$e^1, e^2, \ldots, e^m, u^1 - v^1, u^2 - v^2, \ldots, u^k - v^k \tag{4.2}$$

is acute.

□ Using Definition 4.1 and Theorem 2.1, we conclude that a collection of vectors (4.1) is consistent if and only if there exists a cone relation with an acute convex cone K (without the origin) satisfying the relationships

$$R_+^m \subset K, \quad u^s - v^s \in K, \quad s = 1, 2, \ldots, k. \tag{4.3}$$

Necessity. Let a collection of the pairs of vectors (4.1) be consistent. According to the aforesaid, then there exists an acute convex cone K (without the origin) satisfying (4.3). The differences of the vectors $u^s - v^s, s = 1, 2, \ldots, k$, together with $e^1, e^2, \ldots, e^m$ belong to the cone K and generate a certain convex subcone M of the cone K. A subcone of an acute cone is acute, too; hence, the collection (4.2) generates an acute convex cone.

Sufficiency. Consider the convex cone (without the origin) generated by vectors (4.2). Denote it by M. This cone is acute by the data. Since all unit vectors $e^1, e^2, \ldots, e^m$ belong to M, we have $R^m_+ \subset M$. Therefore, relationships (4.3) for $K = M$ hold for the cone under consideration. ■

A vector $z^* \in R^m$ is called *N-solution* of the system of linear equations $Az = b$ with $A = (a_{ij})_{n\times m}$ and $b \in R^m$ if the equality $Az^* = b$ as well as the inequality $z^* \geq 0_m$ hold.

Theorem 4.2 (algebraic criterion of consistency). *For a collection of the pairs of vectors* (4.1) *to be consistent, a necessary and sufficient condition is that the system of linear homogeneous equations*

$$\sum_{i=1}^{m} \lambda_i e^i + \sum_{s=1}^{k} \mu_s (u^s - v^s) = 0_m \quad (4.4)$$

has no N-solution $\lambda_1, \lambda_2, \ldots, \lambda_m, \mu_1, \mu_2, \ldots, \mu_k$.[1]

□ Necessity. In the contrary, if the system of linear equations (4.4) has the N-solution (λ, μ), where $\lambda = (\lambda_1, \lambda_2, \ldots, \lambda_m), \mu = (\mu_1, \mu_2, \ldots, \mu_k)$, then $-(\lambda, \mu)$ is the N-solution too. This means that the cone generated by the vectors (4.2) is not acute. Due to Theorem 4.1 the collection (4.1) is inconsistent.

Sufficiency. If the collection (4.1) is inconsistent then in according to Theorem 4.1 the cone M generated by the vectors (4.2) is not acute. In this case, there exists nonzero vector $y \in R^m$ such that

$$y = \sum_{i=1}^{m} \lambda_i e^i + \sum_{s=1}^{k} \mu_s (u^s - v^s) \in M, \quad -y = \sum_{i=1}^{m} \lambda'_i e^i + \sum_{s=1}^{k} \mu'_s (u^s - v^s) \in M,$$

where $(\lambda, \mu) \geq 0_{m+k}$ and $(\lambda', \mu') \geq 0_{m+k}$. Since $y + (-y) = 0_m \in M$, we have

$$\sum_{i=1}^{m} (\lambda_i + \lambda'_i) e^i + \sum_{s=1}^{k} (\mu_i + \mu'_s)(u^s - v^s) = 0_m$$

for the vector $(\lambda + \lambda', \mu + \mu') \geq 0_{m+k}$. Thus, this vector is N-solution of the linear system (4.4). ■

Remark 4.1 Combining Theorems 4.1 and 4.2 we obtain that the cone M generated by the vectors (4.2) is acute if and only if the system of homogeneous linear equations (4.4) has no N-solution.

Consider the trivial case of a single information quantum, i.e., $k = 1$. The system of linear equations (4.4) acquires the form

[1]It means that, in these system of equations (if solvable at all), either all numbers $\lambda_1, \lambda_2, \ldots, \lambda_m, \mu_1, \mu_2, \ldots, \mu_k$ are zero or at least one of them is negative.

$$\sum_{i=1}^{m} \lambda_i e^i = -\mu_1 (u^1 - v^1). \tag{4.5}$$

Assume that this system has the N-solution $\lambda_1, \lambda_2, \ldots, \lambda_m, \mu_1$. If $\mu_1 = 0$, then (4.5) becomes the equality $\sum_{i=1}^{m} \lambda_i e^i = 0_m$, where at least one coefficient $\lambda_1, \lambda_2, \ldots, \lambda_m$ is positive. But then this equality is false.

Let $\mu_1 \neq 0$. The vector $u^1 - v^1$ has at least one positive component; therefore, the vector $-\mu_1(u^1 - v^1)$, which appears in the right-hand side of (4.5) and contains at least one negative component, cannot be expressed as the nonnegative linear combination of the unit vectors $e^1, e^2, \ldots, e^m$. Hence, equality (4.5) is again impossible, and the system of linear equations (4.4) has no N-solutions.

Thus, we arrive at the following result.

Corollary 4.1 *If* $k = 1$, *then the pair of vectors inducing a corresponding quantum is consistent. A collection of the pairs of vectors* (4.1) *may be inconsistent only if this collection contains more than one pair.*

For $k = 2$, inconsistency may happen and this fact was demonstrated by Example 4.1. In this example, the system of equations (4.4) has the form

$$\lambda_1 + \mu_1 + \mu_2(-2) = 0,$$

$$\lambda_2 + \mu_1(-3) + \mu_2 = 0.$$

As easily checked, one of the N-solutions is $\lambda_1 = 3,\ \lambda_2 = 1,\ \mu_1 = 1,\ \mu_2 = 2$.

Corollary 4.2 *Information in the form of two quanta stating that criterion i is more important than criterion j with parameters* w_i, w_j *and criterion j is simultaneously more important than criterion i with parameters* w'_j, w'_i *is consistent if and only if* $w_i / w_j > w'_i / w'_j$.

□ It suffices to prove this result for the two-dimensional vectors. Since there is information in the form of the above quanta, we have the relationships $(w_i, -w_j) \succ 0_2$ and $(-w'_i, w'_j) \succ 0_2$. According to Theorem 4.2, this collection is inconsistent if and only if the system of linear equations

$$\lambda_1 + \mu_1 w_i - \mu_2 w'_i = 0,$$
$$\lambda_2 - \mu_1 w_j + \mu_2 w'_j = 0,$$

has the N-solution $\lambda_1, \lambda_2, \mu_1, \mu_2$. Clearly, the equalities $\mu_1 = \mu_2 = 0$ imply the equalities $\lambda_1 = \lambda_2 = 0$, that is impossible. And so, without loss of generality, we may suppose that, e.g., $\mu_2 \neq 0$. In this case, the system of linear equations is equivalent to the inequalities

$$\frac{\mu_1}{\mu_2} \leq \frac{w'_i}{w_i}, \quad \frac{w'_j}{w_j} \leq \frac{\mu_1}{\mu_2},$$

which are by-turn equivalent to $\frac{w'_j}{w_j}, \leq \frac{w'_i}{w_i}$, or $\frac{w_i}{w_j} \leq \frac{w'_i}{w'_j}$. The last inequality is opposite to the desired one $w_i/w_j > w'_i/w'_j$, which must be the case, since in the beginning of the proof we have hypothesized the inconsistency of the collection of the two information quanta. ■

The geometrical meaning of Corollary 4.2 fully manifests itself in the case of the linear criteria. Let $m = 2$, $n = 2$, $f_1(x) = \langle c^1, x\rangle$, and $f_2(x) = \langle c^2, x\rangle$, where $c^1, c^2, x \in R^2$ (see Fig. 4.1).

Since criterion f_1 is more important than criterion f_2 (e.g., with $\theta_{12} = 0.4$), then in the new multicriteria problem (whose Pareto set gives an upper estimate for the desired set of selectable alternatives and vectors) criterion f_2 is replaced by the new second linear criterion with the gradient c^2_{new}. To obtain the end of this vector, we should shift the end of c^2 towards the end of c^1 to the 40% length of the segment between these ends. On the other hand, since criterion f_2 is more important than criterion f_1 (let $\theta_{21} = 0.25$), the new first linear criterion has the gradient c^1_{new} whose end is at the distance of the 25% length of the above segment from the end of the vector c^1 towards the end of the vector c^2. The new vector criterion has the form $(\langle c^1_{new}, x\rangle, \langle c^2_{new}, x\rangle)$. Therefore, by taking into account such information, we observe the mutual change of direction for both gradients, which can be treated as "closer goals."

Corollary 4.3 *Consider two groups of criteria indexes* $i_s \in I$, $j_s \in I$, $s = 1, 2, \ldots, k$, $\{i_1, \ldots, i_k\} \cap \{j_1, \ldots, j_k\} = \emptyset$, *where some (or even all) indexes in the first group (just like in the second group) may be identical. A consistent collection is formed by the pairs of vectors* (4.1) *where each vector* $u^s - v^s$ *has a positive component with index* i_s, *a negative component with index* j_s *and all other components equal to 0,* $s = 1, 2, \ldots, k$.

□ Conjecture the opposite: the system of linear equations (4.4) has the N-solution $\lambda_1, \lambda_2, \ldots, \lambda_m, \mu_1, \mu_2, \ldots, \mu_k$. First, consider the case with at least one positive number among $\mu_1, \mu_2, \ldots, \mu_k$. In this case, the vector $\sum_{s=1}^{k} \mu_s(u^s - v^s)$ contains at least one positive component among the ones belonging to the first group. This contradicts the initial assumption that the sum $\sum_{i=1}^{m} \lambda_i e^i + \sum_{s=1}^{k} \mu_s(u^s - v^s)$ is the zero vector.

If all coefficients $\mu_1, \mu_2, \ldots, \mu_k$ equal 0, the system of equations (4.4) turns into $\sum_{i=1}^{m} \lambda_i e^i = 0_m$, where at least one of the coefficients λ_i is nonzero. But such a system of equations has no nonzero solutions, which contradicts the initial assumption. ■

Using similar considerations as in the proof of Corollary 4.3, we can establish a more general result as follows.

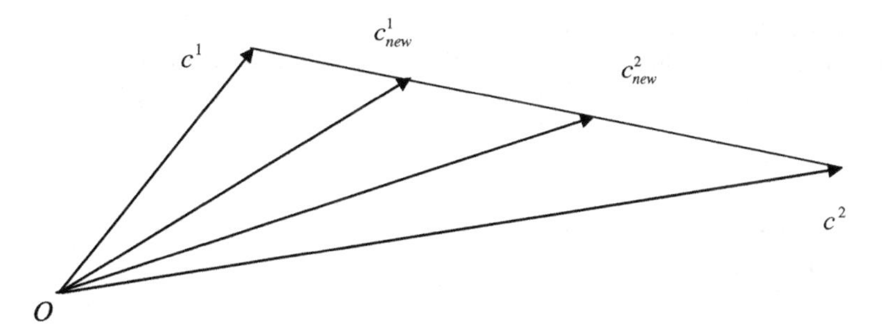

Fig. 4.1 Vectors $c^1, c^2, c^1_{new}, c^2_{new}$

Corollary 4.4 *A collection of the pairs of vectors* (4.1) *is consistent if it satisfies the following conditions. In each vector $u^s - v^s$, all components with the indexes from the set A_s, $A_s \subset I$, are positive, all components with the indexes from the set B_s, $B_s \subset I$, are negative, and all other components are equal to 0, $s = 1, 2, \ldots, k$, and moreover the equality $A_i \cap B_j = \varnothing$ holds for any pair of indexes $i, j \in \{1, 2, \ldots, k\}$.*

Now, formulate another criterion of consistency (more specifically, inconsistency) for a collection of vectors.

Theorem 4.3 (criterion of inconsistency). *For a collection of vectors* (4.1) *to be inconsistent, a necessary and sufficient condition is that in the linear programming problem*

$$\begin{gathered}
\xi_1 + \xi_2 + \ldots + \xi_m + \xi_{m+1} \rightarrow \min, \\
\lambda_1 + \sum_{s=1}^{k} \mu_s (u_1^s - v_1^s) + \xi_1 = 0, \\
\cdots\cdots\cdots\cdots\cdots\cdots\cdots\cdots \\
\lambda_m + \sum_{s=1}^{k} \mu_s (u_m^s - v_m^s) + \xi_m = 0, \\
\lambda_1 + \ldots + \lambda_m + \mu_1 + \ldots + \mu_k + \xi_{m+1} = 1, \\
\lambda_1, \lambda_2, \ldots, \lambda_m \geqq 0;\ \xi_1, \xi_2, \ldots, \xi_m, \xi_{m+1} \geqq 0;\ \mu_1, \mu_2, \ldots, \mu_k \geqq 0,
\end{gathered} \tag{4.6}$$

the optimal value of the goal function is 0.

□ Clearly, the equality constraints in the linear programming problem (4.6) without the artificial variables $\xi_1, \xi_2, \ldots, \xi_{m+1}$ and the equality $\lambda_1 + \ldots + \lambda_m + \mu_1 + \ldots + \mu_k = 1$ coincides with the system of linear equations (4.4). By Theorem 4.3, a collection of the pairs of vectors (4.1) is inconsistent if and only if the system of linear homogeneous equations (4.4) has at least one *N*-solution. By-turn, this holds if and only if the linear programming problem (4.6) has a feasible solution where all artificial variables $\xi_i = 0$, $i = 1, 2, \ldots, m + 1$. The last condition is equivalent to the zero optimal value of the goal function in the linear programming problem (4.6). ■

Remark 4.2 Note that the linear programming problem (4.6) always has an optimal solution, since the values of its goal function are bounded below by 0. Therefore, the optimal value of this function surely exists, being 0 or strictly greater than 0.

4.1.4 Essential Information About Relative Importance of Criteria

We have emphasized earlier that, in practice, the process of information acquisition in the form of quanta is often sequential (at the beginning, one quantum is obtained, then another, and so on). In this case, it is crucial to identify next quanta that contradict the previous ones. Moreover, it is very useful to distinguish between essential and nonessential information. For instance, if we already know that criterion i is more important than criterion j with the degree of compromise 0.5, then a similar message with a smaller coefficient contains nothing new (essential) in comparison with the previous message and can be therefore ignored.

Consider a given consistent collection of the pairs of vectors (4.1). Supplement this collection with another pair of vectors u^{k+1}, v^{k+1} such that $u^{k+1} - v^{k+1} \in N^m$. As a result, we obtain the "expanded" collection of the pairs of vectors

$$u^i, v^i \in R^m, \quad u^i - v^i \in N^m, \quad i = 1, 2, \ldots, k+1. \tag{4.7}$$

Definition 4.2 For a consistent collection of the pairs of vectors (4.1), a pair u^{k+1}, v^{k+1} is called *essential* if the convex cone generated by the unit vectors $e^1, e^2, \ldots, e^m$ together with the vectors $u^i - v^i, i = 1, 2, \ldots, k+1$, does not coincide with the convex cone generated by the same unit vectors and the vectors $u^i - v^i, i = 1, 2, \ldots, k$.

The meaning of this definition is that essential additional information about the DM's preference relation must modify the existing cone preference relation. Obviously, the non-coincident cones defined by collections (4.1) and (4.7) may occur only if the cone generated by the expanded collection is wider than the one generated by the original collection of k vectors.

Theorem 4.4 (criterion of consistency and essentiality). *Let a collection of the pairs of vectors* (4.1) *be consistent. For the expanded collection* (4.7) *to be consistent and simultaneously the pair of vectors* u^{k+1}, v^{k+1} *to be essential, a necessary and sufficient condition is that the two systems of linear inhomogeneous equations*

$$\sum_{i=1}^{m} \lambda_i e^i + \sum_{s=1}^{k} \mu_s (u^s - v^s) = \pm (u^{k+1} - v^{k+1}) \tag{4.8}$$

have no nonnegative solution $\lambda_1, \lambda_2, \ldots, \lambda_m, \mu_1, \mu_2, \ldots, \mu_k$. *Here one system of equations corresponds to* $+$ *and the other to* $-$.

□ First, we deal with the issue of consistency. According to the algebraic criterion of consistency, the expanded collection of vectors (4.7) is consistent if and only if the system of linear homogeneous equations

$$\sum_{i=1}^{m} \lambda_i e^i + \sum_{s=1}^{k} \mu_s (u^s - v^s) + \mu_{k+1} (u^{k+1} - v^{k+1}) = 0_m \tag{4.9}$$

has no N-solution. Check that this condition is equivalent to the absence of nonnegative solutions for the system of equations (4.8–) (the system of equations (4.8) with the minus sign in the right-hand side). Really, if the system of equations (4.9) has no N-solution, then the system of equations (4.8–) is not solvable in the nonnegative variables. Conversely, if the second system of the linear equations [i.e., (4.8–)] has no nonnegative solutions, while the first system has the N-solution $\lambda_1, \lambda_2, \ldots, \lambda_m, \mu_1, \mu_2, \ldots, \mu_k$, we arrive at contradiction. Indeed, the case $\mu_{k+1} = 0$ is impossible, since the collection of vectors (4.1) is consistent; hence, $\mu_{k+1} > 0$. Then, by dividing both sides of (4.9) by μ_{k+1}, we establish that the system of equations (4.8−) has a nonnegative solution. This result contradicts the initial assumption. The first part of Theorem 4.4 (dedicated to consistency) is proved.

Now, we argue the second part associated with the relevance of the pair of vectors u^{k+1}, v^{k+1}. According to Definition 4.2, this pair of vectors is essential if and only if the vector $u^{k+1} - v^{k+1}$ does not belong to the convex cone generated by the vectors $e^1, e^2, \ldots, e^m, u^1 - v^1, u^2 - v^2, \ldots, u^k - v^k$. The last holds if and only if the system of linear homogeneous equations (4.8+) has no nonnegative solution. ■

Remark 4.3 Theorem 4.4 consists of two parts, one relating to the consistency of the pair of vectors u^{k+1}, v^{k+1} and the other to their essence. As follows from the proof, the first part of the theorem concerns the existence of a nonnegative solution to the system of equations (4.8−), while the issue of essence is settled in terms of the solutions to the system of equations (4.8+).

4.2 Consideration of Two Elementary Information Quanta

4.2.1 Case of Two Mutually Independent Quanta

Consider four given non-empty collections of criteria indexes A_1, B_1, A_2, and B_2 such that $A_1 \cap B_1 = \varnothing$ and $A_2 \cap B_2 = \varnothing$. Suppose that the group of criteria A_1 is more important than the group B_1 with definite collections of the parameters and, simultaneously, the group of criteria A_2 is more important than the group B_2 with the corresponding collections of the parameters. In other words, there are two information quanta. We say that these quanta are *mutually independent* if

$$A_1 \cap A_2 = \varnothing, B_1 \cap B_2 = \varnothing, A_1 \cap B_2 = \varnothing, A_2 \cap B_1 = \varnothing.$$

In a similar fashion, one may introduce the notion of mutual independence for an arbitrary finite number of information quanta, requiring the pairwise disjointness of all groups of criteria involved.

For Pareto set reduction using two mutually independent elementary information quanta, it is necessary to apply Theorem 3.5 twice, which gives the recalculation formulas for the vector criterion. In the beginning, this theorem can be applied, e.g., for taking into account the first quantum (i.e., to the groups of criteria A_1 and B_1). As a result, instead of the criteria of the less important group B_1, we calculate the new criteria by formula (3.5). Next, this theorem is applied to the second group of criteria A_2 and B_2, which gives the recalculated criteria of the group B_2 using the same formula (3.5). The described procedure yields the new vector criterion, in terms of which the set of Pareto optimal alternatives (vectors) represents an upper estimate for the unknown set of selectable alternatives $C(X)$ (selectable vectors $C(Y)$).

Now, consider situation where criterion i is more important than criterion j and, by-turn, the latter criterion is more important than criterion k, $i \neq j$, $j \neq k$, $i \neq k$. Here we also deal with two elementary information quanta, but they are not mutually independent. Nevertheless, Theorem 2.5 can be applied twice for taking this collection into account and constructing the new vector criterion. Recall that this theorem concerns consideration of an elementary information quantum. First, one should recalculate criterion k for using information that criterion j is more important than criterion k. And second, one should recalculate criterion j for using information that criterion i is more important than criterion j. As a result, one obtains the new vector criterion where all components (except j and k) remain the same. The set of Pareto optimal alternatives (Pareto optimal vectors) in terms of the new vector criterion is the refined upper estimate for the unknown set of selectable alternatives (selectable vectors) that corresponds to these information quanta.

4.2.2 Case Where One of Two Criteria Is More Important Than the Two Others

If several quanta are involved for Pareto set reduction, one should have the following aspect in mind. Let criterion i be more important than criterion j with the degree of compromise θ_{ij} and, in addition, let criterion i be more important than criterion k $(k \neq j)$ with the degree of compromise θ_{ik}. Thereby, there is a given collection of these two elementary information quanta, and such a situation resembles the one where criterion i is more important than the group of criteria $\{j, k\}$ with the degrees of compromise θ_{ij} and θ_{ik}.

By virtue of Theorem 3.1, *if criterion i is more important than the group of criteria* $\{j,k\}$ *with the degrees of compromise* θ_{ij} *and* θ_{ik}*, respectively, then criterion i is more important than each of criteria j and k separately with the same degrees of compromise.*

Now, let criterion i be more important than criteria j and k separately with the degrees of compromise θ_{ij} and θ_{ik}. In this case, the DM is willing to lose separately the quantity w_j^* in terms of criterion j or the quantity w_k^* in terms of criterion k for gaining the quantity w_i^* in terms of criterion i. As easily seen, this does not imply that the DM agrees to lose simultaneously the quantities w_j^* and w_k^* in terms of criteria j and k, respectively, as a compensation for gaining the quantity w_i^* in terms of criterion i. In other words, *if criterion i is more important than each of criteria j and k separately with the degrees of compromise* θ_{ij} *and* θ_{ik}*, respectively, then generally criterion i is not more important than the group of criteria* $\{j,k\}$ *with the same degrees of compromise.*

However, note that *the higher importance of criterion i against each of criteria j and k separately with the degrees of compromise* θ_{ij} *and* θ_{ik}*, respectively, implies the higher importance of criterion i against the group of criteria* $\{j,k\}$*, but with smaller degrees of compromise.*

□ Really, the termwise addition of the relationships $y' \succ 0_m$ and $y'' \succ 0_m$, where the vectors y' and y'' are defined by the equalities

$$y_i' = w_i^*, \quad y_j' = -w_j^*, \quad y_s' = 0 \text{ for all } s \in I \backslash \{i,j\},$$

$$y_i'' = \bar{w}_i, \quad y_k'' = -\bar{w}_k, \quad y_s'' = 0 \text{ for all } s \in I \backslash \{i,k\},$$

yields the relationship $y = y' + y'' \succ 0_m$, where the vector y has the components

$$y_i = w_i^* + \bar{w}_i, \quad y_j = -w_j^*, \quad y_k = -\bar{w}_k, \quad y_s = 0 \text{for all} s \in I \backslash \{i,j,k\}.$$

This means that criterion i is more important than the group of criteria $\{j,k\}$ with the degrees of compromise

$$\theta_{ij}' = \frac{w_j^*}{w_i^* + \bar{w}_i + w_j^*} < \theta_{ij}, \quad \theta_{ik}' = \frac{\bar{w}_k}{w_i^* + \bar{w}_i + \bar{w}_k} < \theta_{ik}.$$

■

The next result shows how the vector criterion should be recalculated for taking into account a collection of the two information quanta.

Theorem 4.5 (in terms of vectors). *Assume that there are given two information quanta, one stating that criterion i is more important than criterion j with parameters* w_i, w_j *and the other that criterion i is more important than criterion k with parameters* w_i', w_k'*. Then for any set of selectable vectors* $C(Y)$ *we have the inclusions*

$$C(Y) \subset \hat{P}(Y) \subset P(Y), \tag{4.10}$$

where $\hat{P}(Y) = f(P_g(X))$ *is the set of feasible vectors corresponding to the set of Pareto optimal alternatives in the multicriteria problem with the initial set X and the new* $(m+1)$*-dimensional vector criterion g with the components*

$$\begin{aligned} g_j &= w_j f_i + w_i f_j, \quad g_k = w'_k f_i + w'_i f_k, \\ g_{m+1} &= w_j w'_k f_i + w_i w'_k f_j + w'_i w_j f_k, \\ g_s &= f_s \quad \text{for all } s \in I \backslash \{j, k\}. \end{aligned} \tag{4.11}$$

□ I. Denote by K the acute convex cone (without the origin) of the cone relation $\succ$.

According to Definition 2.4, the existence of these information quanta means that the relationships $y' \succ 0_m$ and $y'' \succ 0_m$ hold, which is equivalent to the inclusions $y' \in K$ and $y'' \in K$ for the vectors y' and y'' with the components

$$y'_i = w_i, \quad y'_j = -w_j, \quad y'_s = 0 \quad \text{for all } s \in I \backslash \{i, j\},$$

$$y''_i = w'_i, \quad y''_k = -w'_k, \quad y''_s = 0 \text{ for all } s \in I \backslash \{i, k\}.$$

Due to Corollary 4.3 given collection of two information quanta is consistent. Let M be the acute convex cone (without the origin) generated by the vectors $e^1, \ldots, e^m, y', y''$. This cone is generated by the same collection without the vector e^i, since the latter can be expressed as N-combination of the vectors e^j, y'. Therefore, the cone M coincides with the set of all N-combinations of the form

$$\lambda_1 e^1 + \ldots + \lambda_{i-1} e^{i-1} + \lambda'_i y' + \lambda''_i y'' + \lambda_{i+1} e^{i+1} + \ldots + \lambda_m e^m.$$

II. Now, we demonstrate that the cone M coincides with the set of nonzero solutions to the system of linear homogeneous inequalities

$$\begin{aligned} & y_s \geqq 0 \text{ for all } s \in I \backslash \{j, k\}, \\ & w_j y_i + w_i y_j \geqq 0, \\ & w'_k y_i + w'_i y_k \geqq 0, \\ & w_j w'_k y_i + w_i w'_k y_j + w'_i w_j y_k \geqq 0. \end{aligned} \tag{4.12}$$

To this end, find the fundamental system of solutions of this system of inequalities by considering the corresponding system of linear equations

$$\begin{aligned} \langle e^s, y \rangle &= 0 \quad \text{for all } s \in I \backslash \{j, k\}, \\ \langle \bar{y}', y \rangle &= 0, \\ \langle \bar{y}'', y \rangle &= 0, \\ \langle \bar{y}, y \rangle &= 0, \end{aligned} \tag{4.13}$$

where the components of the vectors $\bar{y}', \bar{y}'', \bar{y}$ are defined by the equalities

$$\bar{y}'_i = w_j, \bar{y}'_j = w_i, \bar{y}'_s = 0 \quad \text{for all } s \in I \backslash \{i,j\},$$

$$\bar{y}''_i = w'_k, \bar{y}''_k = w'_i, \bar{y}''_s = 0 \quad \text{for all } s \in I \backslash \{i,k\},$$

$$\bar{y}_i = w_j w'_k, \bar{y}_j = w_i w'_k, \bar{y}_k = w'_i w_j, \bar{y}_s = 0 \quad \text{for all } s \in I \backslash \{i,j,k\}.$$

System (4.13) contains $(m + 1)$ linear equations, and any subset of $(m - 1)$ vectors from the collection e^s for all $s \in I \backslash \{j,k\}$, $\bar{y}', \bar{y}'', \bar{y}$ that form this system of linear equations is linear independent. Hence, to construct the fundamental system of solutions to the system of linear inequalities (4.12), it suffices to find the one-dimensional nonzero solutions to all possible subsystems of (4.13) resulting from (4.13) after elimination of some two equations. The solutions obtained by this approach must satisfy the system of inequalities (4.12).

We start to remove sequentially two equations from system (4.13). First, consider the case where each pair of the removed equations includes the last equation. After elimination of the last two equations, we obtain the subsystem with the nonzero solution e^k. By removing the last equation together with equation (m–1), we get the subsystem with the solution e^j. Elimination from (4.13) the last equation together with one of the equations $\langle e^s, y \rangle = 0$ for all $s \neq i$ yields the subsystem that has the solution e^s. If the last equation is removed together with $\langle e^i, y \rangle = 0$, then the resulting subsystem has no nonzero solution satisfying the system of inequalities (4.12).

Now, examine the case where system (4.13) is truncated by eliminating a pair of equations that contains equation $(m - 1)$. If we remove the preceding equation together with equation $(m - 1)$, the resulting subsystem has no nonzero solutions satisfying the system of inequalities (4.12). After elimination of equation $(m - 1)$ together with another equation of the form $\langle e^s, y \rangle = 0$ for $s \neq i$, we obtain the subsystem with the solution e^s. And elimination of equation (m–1) together with the equation $\langle e^i, y \rangle = 0$ yields the subsystem with the solution y'.

Similarly, one may analyze the case where equation $(m - 1)$ is eliminated together with another equation of the form $\langle e^s, y \rangle = 0$. This gives one more nonzero solution y'' for $s = i$ that satisfies the system of linear inequalities (4.12).

Note that no new solutions are obtained as the result of removing two equations of the form $\langle e^s, y \rangle = 0$ from system (4.13).

We have found the collection of vectors $e^1, \ldots, e^{i-1}, y', y'', e^{i+1}, \ldots, e^m$ generating the cone of solutions to the system of linear inequalities (4.12). This collection coincides with the collection of vectors generating the cone M. Thereby, it is shown that the set of nonzero solutions to the system of linear inequalities (4.12) coincides with the cone M.

III. The inclusions $R^m_+ \subset M \subset K$ obviously imply the inclusions

$$\operatorname{Ndom} Y \subset \hat{P}(Y) \subset P(Y), \tag{4.14}$$

where

$\hat{P}(Y) = \{y^* \in Y \mid \text{there exists no } y \in Y \text{ such that } y - y^* \in M\}$
represents the set of nondominated elements of the set Y ordered by the cone relation with the cone M.

Choose arbitrary two vectors $y = f(x)$, $y^* = f(x^*)$, $f(x) \neq f(x^*)$ for $x, x^* \in X$. Owing to the established coincidence of the solution set of the system of linear inequalities (4.12) and the cone M (see part II), the inclusion $f(x) - f(x^*) \in M$ takes place if and only if

$$f_s(x) - f_s(x^*) \geqq 0 \quad \text{for all } s \in I \backslash \{j, k\},$$

$$w_j(f_i(x) - f_i(x^*)) + w_i(f_j(x) - f_j(x^*)) \geqq 0,$$

$$w'_k(f_i(x) - f_i(x^*)) + w'_i(f_k(x) - f_k(x^*)) \geqq 0,$$

$$w_j w'_k(f_i(x) - f_i(x^*)) + w_i w'_k(f_j(x) - f_j(x^*)) + w'_i w_j(f_k(x) - f_k(x^*)) \geqq 0,$$

where at least one inequality is strict. Clearly, these inequalities can be rewritten as $g(x) \geq g(x^*)$ in terms of the vector function g defined by formulas (4.11). Hence, it appears that $\hat{P}(Y)$ is the set of vectors answering the set of Pareto optimal alternatives in the multicriteria problem with the initial set X and the new vector criterion g, i.e., $\hat{P}(Y) = f(P_g(X))$.

IV. To finish the proof, just take into account the inclusion $C(Y) \subset \operatorname{Ndom}(Y)$, which holds for any set of selectable vectors $C(Y)$. ■

Proved Theorem 4.5 can be easily reformulated in terms of alternatives as follows.

Theorem 4.5 (in terms of alternatives). *Assume that there are two information quanta, one stating that criterion i is more important than criterion j with parameters w_i, w_j and the other that criterion i is more important than criterion k with parameters w'_i, w'_k. Then for any set of selectable alternatives we have the inclusions*

$$C(X) \subset P_g(X) \subset P_f(X), \tag{4.15}$$

where $P_g(X)$ is the set of Pareto optimal alternatives in the multicriteria problem with the initial set X and the new $(m+1)$–dimensional g defined by (4.11).

Let us give a geometrical illustration for Theorem 4.5 in the case of linear criteria. Let $m = n = 3$, $f_i(x) = \langle c^i, x \rangle$, $i = 1, 2, 3$, where $c^1, c^2, c^3, x \in R^3$ (see Fig. 4.2). Suppose that criterion f_1 is more important than criterion f_2 with the parameters $w_1 = 1$, $w_2 = 3$ and also more important than criterion f_3 with the parameters $w'_1 = 2, w_3 = 3$. By taking into account the higher importance of criterion f_1 against criterion f_2, we arrive at the tricriteria problem with the vectors c^1, c^2_{new}, c^3 representing the gradients of the three linear criteria. Similarly, the consideration of the higher importance of criterion f_1 against criterion f_3 yields the tricriteria problem with the three vectors c^1, c^2, c^3_{new}.

Thereby, we obtain the two cones of goals associated with the two information quanta available. For the simultaneous consideration of both quanta, it is necessary to intersect these cones, which gives the cone generated by the four vectors $c^1, c^2_{new}, c^3_{new}, c^4$. The latter represents the cone of goals in the new problem.

The next result shows that Theorem 4.5 can be used for any criteria measured in quantitative scales.

Theorem 4.6 *Inclusions* (4.10) *and* (4.15) *are invariant with respect to a linear positive transformation applied to the components of the vector criterion g defined by equalities* (4.11).

□ Based on the proof of Theorem 2.7, it suffices to verify the invariance of the sets $\hat{P}(Y)$ and $P_g(X)$ from (4.10) and (4.15) with respect to a linear positive transformation of the last criterion g_{m+1} only.

We have

$$g_{m+1} = w_j w'_k y_i + w_i w'_k y_j + w_j w'_i y_k.$$

Replace y_s with $\tilde{y}_s = \alpha_s y_s + c_s (\alpha_s > 0), \quad s = i, j, k$, in the formula of the criterion g_{m+1}. We get

$$\begin{aligned}\tilde{g}_{m+1} &= \alpha_j \alpha_k w_j w'_k (\alpha_i y_i + c_i) + \alpha_i \alpha_k w_i w'_k (\alpha_j y_j + c_j) + \alpha_i \alpha_j w_j w'_i (\alpha_k y_k + c_k) \\ &= \alpha_i \alpha_j \alpha_k w_j w'_k y_i + \alpha_i \alpha_j \alpha_k w_i w'_k y_j + \alpha_i \alpha_j \alpha_k w_j w'_i y_k + C\end{aligned}$$

where the constant C is independent of y_i, y_j, y_k.

Clearly, the transformed criterion $\tilde{g}_{m+1}$ can be constructed from g_{m+1} via its multiplication by the positive number $\alpha_i \alpha_j \alpha_k$ and addition of the constant C. Conversely, by deducting the constant C from $\tilde{g}_{m+1}$ and dividing the result by the number $\alpha_i \alpha_j \alpha_k$, we obtain g_{m+1}. Hence, it follows that the following two strict inequalities

$$g_{m+1} = w_j w'_k y_i + w_i w'_k y_j + w_j w'_i y_k > w_j w'_k y^*_i + w_i w'_k y^*_j + w_j w'_i y^*_k = g^*_{m+1}$$

and

$$\tilde{g}_{m+1} = w_j w'_k \tilde{y}_i + w_i w'_k \tilde{y}_j + w_j w'_i \tilde{y}_k > w_j w'_k \tilde{y}^*_i + w_i w'_k \tilde{y}^*_j + w_j w'_i \tilde{y}^*_k = \tilde{g}^*_{m+1}$$

for the criterion g_{m+1} and transformed criterion $\tilde{g}_{m+1}$, respectively, are equivalent to each other. Subsequently, inclusions (4.10) and (4.15) are invariant with respect to a linear positive transformation of the criterion g_{m+1}, ergo all components of the vector function g. ■

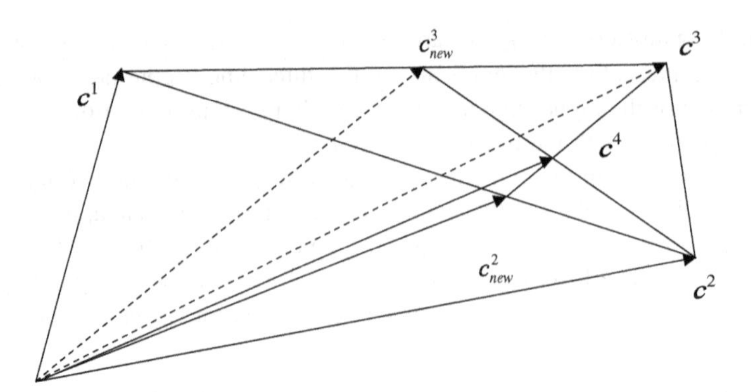

Fig. 4.2 Vectors $c^1, c^2, c^3, c^4, c^2_{new}, c^3_{new}$.

4.2.3 *Pareto Set Reduction When Each of Two Criteria Is More Important Than the Third One*

This subsection considers the case of two information quanta, one stating that criterion i is more important than criterion k with parameters w_i, w_k and the other that criterion j is more important than criterion k with parameters w'_j, w'_k.

Prior to detailed analysis, we compare this case with the existence of one quantum about the higher importance of the group of criteria $\{i,j\}$ against criterion k, under the same degrees of compromise in both cases. According to Theorem 3.1, *if each (or at least one) of criteria i and j separately is more important than criterion k, then the group of criteria $\{i,j\}$ is more important than criterion k with the same degrees of compromise, but the converse fails.* And so, the two quanta have higher relevance of information than the one quantum stating that the group of two criteria is more important than the third criterion.

The above-mentioned pair of information quanta should be used on the basis of the following theorem.

Theorem 4.7 *Assume that there are two information quanta, one stating that criterion i is more important than criterion k with parameters w_i, w_k and the other that criterion j is more important than criterion k with parameters w'_j, w'_k. Then for any set of selectable vectors $C(Y)$ we have inclusions (4.10), where $\hat{P}(Y) = f(P_g(X))$ is the set of vectors corresponding to the set of Pareto optimal alternatives in the multicriteria problem with the initial set X and the vector criterion g with the components*

$$\begin{aligned} g_s &= f_s \quad \textit{for all } s \in I \backslash \{k\}, \\ g_k &= w_k w'_j f_i + w_i w'_k f_j + w_i w'_j f_k. \end{aligned} \tag{4.16}$$

□ I. Let K be the acute convex cone (without the origin) of the cone relation $\succ$.

By Definition 2.4, we have the relationships $y' \succ 0_m$ and $y'' \succ 0_m$ for the vectors y' and y'' with the components

$$y'_i = w_i, \quad y'_i = w_i, \quad , y'_k = -w_k, \quad y'_s = 0 \quad \text{for all } s \in I \backslash \{i, k\},$$

$$y''_j = w'_j, \quad y''_j = w''_j, \quad , y''_k = -w''_k, \quad y''_s = 0 \quad \text{for all } s \in I \backslash \{j, k\}.$$

Due to Corollary 4.3 given collection of two information quanta is consistent. Let M be the acute convex cone (without the origin) generated by the vectors $e^1, e^2, \ldots, e^m, y', y''$. The vectors e^i and e^j can be represented as the linear positive combinations of the vectors e^k, y' and e^k, y'', respectively. Hence, the cone M is generated by the collection of vectors

$$e^1, \ldots, e^{i-1}, y', e^{i+1}, \ldots, e^{j-1}, y'', e^{j+1}, \ldots, e^m, \tag{4.17}$$

which means that it coincides with the set of all N-combinations

$$\lambda_1 e^1 + \ldots + \lambda_{i-1} e^{i-1} + \lambda_i y' + \lambda_{i+1} e^{i+1} + \ldots + \lambda_{j-1} e^{j-1} + \lambda_j y'' + \lambda_{j+1} e^{j+1} + \ldots + \lambda_m e^m.$$

II. Now, prove that the cone M coincides with the set of nonzero solutions to the system of linear homogenous inequalities

$$\begin{aligned} & y_s \geqq 0 \quad \text{for all } s \in I \backslash \{k\}, \\ & w_k w'_j y_i + w_i w'_k y_j + w_i w'_j y_k \geqq 0. \end{aligned} \tag{4.18}$$

To this end, find the fundamental system of solutions of (4.18) by considering the corresponding system of linear equations

$$\begin{aligned} & \langle e^s, y \rangle = 0 \quad \text{for all } s \in I \backslash \{k\}, \\ & \langle \bar{y}, y \rangle = 0, \end{aligned} \tag{4.19}$$

where $\bar{y}_i = w_k w'_j$, $\bar{y}_j = w_i w'_k$, $\bar{y}_k = w_i w'_j$, and $\bar{y}_s = 0$ for all $s \in I \backslash \{i, j, k\}$.

System (4.19) contains m equations. Each subsystem of $(m - 1)$ vectors from the system $e^1, \ldots, e^{k-1}, \bar{y}, e^{k+1}, \ldots, e^m$ is linearly independent. Therefore, to construct the fundamental system of solutions to the system of inequalities (4.18), it suffices to find a nonzero solution to each of the subsystems resulting from (4.19) after elimination of a certain equation. Note that the desired solution must satisfy the system of inequalities (4.18).

By removing the last equation from system (4.19), we get the subsystem with the solution e^j. Elimination from (4.19) the equation $\langle e^s, y \rangle = 0$ for $s = i$ (or $s = j$) yields the subsystem that has the solution y' (or y''). If the equation $\langle e^s, y \rangle = 0$ for $s \in I \backslash \{i, j, k\}$ is removed, then the resulting subsystem has the solution e^s.

Thus, one of the fundamental systems of solutions to the system of linear inequalities (4.18) has form (4.17). Hence, the cone M coincides with the set of all nonzero solutions to the system of linear inequalities (4.18).

III. The inclusions

$$R_+^m \subset M \subset K$$

imply

$$\operatorname{Ndom} Y \subset \hat{P}(Y) \subset P(Y), \tag{4.20}$$

where

$\hat{P}(Y) = \{y^* \in Y|$ there exists no $y \in Y$ such that $y - y^* \in M\}$
represents the set of nondominated elements of the set Y in terms of the cone relation with the cone M.

Choose two arbitrary elements $x, x^* \in X, y = f(x), y^* = f(x^*), f(x) \neq f(x^*)$. Owing to part II the inclusion $f(x) - f(x^*) \in M$ takes place if and only if the vector $f(x) - f(x^*)$ satisfies the inequalities

$$f_s(x) - f_s(x^*) \geqq 0 \quad \text{for all } s \in I \backslash \{k\},$$

$$w_k w_j'(f_i(x) - f_i(x^*)) + w_i w_k'(f_j(x) - f_j(x^*)) + w_i w_j'(f_k(x) - f_k(x^*)) \geqq 0,$$

where at least one inequality is strict. These inequalities can be rewritten as $g(x) \geq g(x^*)$ in terms of the vector function g defined by formulas (4.16). Hence, $\hat{P}(Y) = f(P_g(X))$.

IV. To finish the proof, just take into account the inclusions (4.20) and $C(Y) \subset \operatorname{Ndom}(Y)$, which holds for any set $C(Y)$. ■

For the geometrical illustration of Theorem 4.7, consider the multicriteria choice problem with the linear criteria, where $m = n = 3$, $f_s(x) = \langle c^s, x \rangle$, $s = 1, 2, 3$, $w_1 = 4$, $w_3 = 1$, and $w_2' = 2$, $w_3' = 3$ (see Fig. 4.3). The cone of goals associated with the new multicriteria problem has three facets and is generated by the vectors c^1, c^2, c^3_{new}. This cone is the intersection of two three-facet cones, one corresponding to the multicriteria choice problem with the quantum that criterion f_1 is more important than criterion f_3 and the other to the multicriteria choice problem with the quantum that criterion f_2 is more important than criterion f_3.

Theorem 4.7 can be used for any multicriteria choice problem with vector function f whose components are measured in quantitative scales, as shown by the next result.

Theorem 4.8 *Inclusions* (4.10) *and* (4.15) *are invariant with respect to a linear positive transformation applied to the components of the vector criterion g defined by equalities* (4.16).

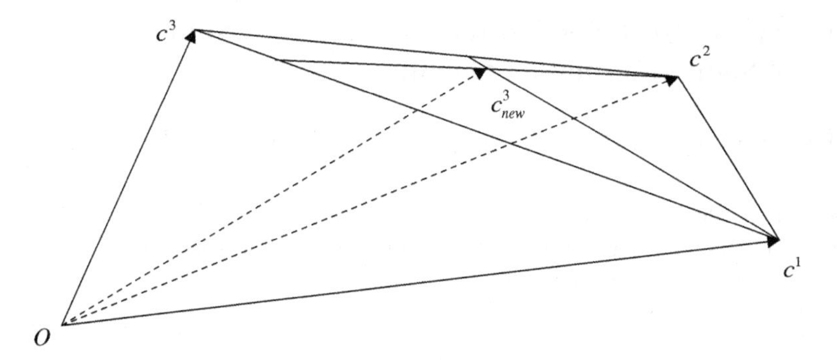

Fig. 4.3 Vectors c^1, c^2, c^3, c^3_{new}.

□ Just like in the proof of Theorem 4.4, it suffices to verify the invariance of the sets $\hat{P}(Y)$ and $P_g(X)$ from (4.10) and (4.15) with respect to a linear positive transformation of the criterion g_k only.

We have

$$g_k = w_k w'_j y_i + w_i w'_k y_j + w_i w'_j y_k.$$

Here replace y_s with $\tilde{y}_s = \alpha_s y_s + c_s (\alpha_s > 0), \quad s = i, j, k$. We get

$$\begin{aligned}\tilde{g}_k &= \alpha_k \alpha_j w_k w'_j (\alpha_i y_i + c_i) + \alpha_i \alpha_k w_i w'_k (\alpha_j y_j + c_j) + \alpha_i \alpha_j w_i w'_j (\alpha_k y_k + c_k) \\ &= \alpha_i \alpha_j \alpha_k w_k w'_j y_i + \alpha_i \alpha_j \alpha_k w_i w'_k y_j + \alpha_i \alpha_j \alpha_k w_i w'_j y_k + C\end{aligned}$$

where the constant C is independent of y_i, y_j, y_k.

According to the last expression, the transformed criterion $\tilde{g}_k$ can be constructed from g_k via its multiplication by the positive number $\alpha_i \alpha_j \alpha_k$ and addition of the constant C. Conversely, by deducting the constant C from $\tilde{g}_k$ and dividing the result by the number $\alpha_i \alpha_j \alpha_k$, we easily obtain g_k. Hence, it follows that the strict inequalities

$$g_k = w_k w'_j y_i + w_i w'_k y_j + w_i w'_j y_k > w_k w'_j y^*_i + w_i w'_k y^*_j + w_i w'_j y^*_k = g^*_k$$

and

$$\tilde{g}_k = w_k w'_j \tilde{y}_i + w_i w'_k \tilde{y}_j + w_i w'_j \tilde{y}_k > w_k w'_j \tilde{y}^*_i + w_i w'_k \tilde{y}^*_j + w_i w'_j \tilde{y}^*_k = \tilde{g}^*_k$$

for the criterion g_k and transformed criterion $\tilde{g}_k$, respectively, are equivalent to each other. Subsequently, inclusions (4.10) and (4.15) are invariant with respect to a linear positive transformation of the criterion g_k. ■

4.3 Pareto Set Reduction Based on a Finite Number of Some Information Quanta

4.3.1 Usage of Information Quanta of Point-Set Type

Theorem 4.5 deals with the case where one criterion is more important than each of the two others. In what follows, we consider the general case where a group of criteria is more important than other two groups of criteria.

Theorem 4.9 *Assume that there are three given pairwise disjoint subsets of criteria indexes* $A, B, C \subset I$ *and information from the DM stating that* $y' \succ 0_m$ *and* $y'' \succ 0_m$, *where the vectors* y' *and* y'' *have the components*

$$y_i' = w_i' \quad \text{for all} \quad i \in A, \quad y_j' = -w_j' \quad \text{for all}\, j \in B, \quad y_s' = 0 \quad \text{for all}\, s \in I \backslash (A \cup B),$$

$$y_i'' = w_i'' \quad \text{for all} \quad i \in A, \quad y_k'' = -w_k'' \quad \text{for all}\, k \in C, \quad y_s'' = 0 \quad \text{for all}\, s \in I \backslash (A \cup C),$$

where w_i', w_j', w_i'', w_j'' *are fixed positive numbers.*

Then this information is consistent, and for any set of selectable vectors $C(Y)$ *we have inclusions* (4.10), *where* $\hat{P}(Y) = f(P_g(X))$ *gives the set of feasible vectors corresponding to the set of Pareto optimal alternatives in the multicriteria problem with the initial set* X *and the new* p-*dimensional vector criterion* g, $p = m - |B| - |C| + |A| \cdot |B| + |A| \cdot |C| + |A| \cdot |B| \cdot |C|$, *that consists of the components*

$$\begin{gathered} g_{ij}' = w_j' f_i + w_i' f_j \quad \text{for all}\, i \in A, j \in B, \\ g_{ik}'' = w_k'' f_i + w_i'' f_k \quad \text{for all}\, i \in A, k \in C, \\ g_{ijk} = w_j' w_k'' f_i + w_i' w_k'' f_j + w_j' w_i'' f_k \quad \text{for all}\, i \in A, j \in B, k \in C, \\ g_s = f_s \quad \text{for all}\, s \in I \backslash (B \cup C). \end{gathered} \tag{4.21}$$

□ This result is proved using a standard scheme.

I. At the beginning, check the consistency of the available information. By Theorem 4.2, this information is consistent if and only if the system of linear algebraic equations

$$\sum_{i=1}^{m} \lambda_i e^i + \lambda_{m+1} y' + \lambda_{m+2} y'' = 0_m \tag{4.22}$$

has no N-solution for $\lambda_1, \ldots, \lambda_{m+2}$. Here e^i means the m-dimensional vector with 1 as component i and 0's as the other components. System (4.22) can be rewritten in the expanded form

$$\lambda_i + \lambda_{m+1} w_i' + \lambda_{m+2} w_i'' = 0 \quad \text{for all } i \in A,$$

$$\lambda_j - \lambda_{m+1} w_j' = 0 \quad \text{for all } j \in B,$$

$$\lambda_k - \lambda_{m+2} w_k'' = 0 \quad \text{for all } k \in C,$$

$$\lambda_l = 0 \quad \text{for all } l \in I \backslash (A \cup B \cup C).$$

The equations in the first row imply the equalities $\lambda_i = \lambda_{m+1} = \lambda_{m+2} = 0$ for all $i \in A$. In this case, the other equations yield the equalities $\lambda_j = 0$ for the other indexes $j \in I \backslash A$. Hence, system (4.22) has the trivial (zero) solution as the only nonnegative solution. This establishes the consistency of the existing information in the form of the relationships $y' \succ 0_m$ and $y'' \succ 0_m$.

II. Recall that K is the acute convex cone (without the origin) of the cone relation $\succ$. Denote by M the acute convex cone (without the origin) generated by the vectors $e^1, \ldots, e^m, y', y''$.

Here two cases are possible, namely, $|A| > 1$ and $|A| = 1$. In the former case, the generators of the cone M are all the vectors $e^1, e^2, \ldots, e^m, y', y''$, since none of them can be expressed as N-combination of the other vectors from this collection (due to the acuteness of the cone M). In the latter case (i.e., if $A = \{i\}$), the vector e^i can be represented by the linear positive combination of the vector y' and all vectors e^s for $s \in B$. And so, in this case, the generators of the cone M are $e^1, e^2, \ldots, e^m, y', y''$ without the vector e^i. We will first analyze the case $|A| > 1$ and then the other.

For the polyhedral cone M, introduce the dual cone D (without the origin) of the form

$$D = \{y \in R^m | \langle u, y \rangle \geqq 0 \text{ for all } u \in M\} \backslash \{0_m\}.$$

The generators of the polyhedral cone D are the inner normals to the $(m-1)$-dimensional facets of the cone M, and conversely, the generators of the cone M are the inner normals to the $(m-1)$-dimensional facets of the cone D, see [62].

Since the cone M is generated by the vectors $e^1, e^2, \ldots, e^m, y', y''$, the set of nonzero solutions to the system of linear homogenous inequalities

$$\begin{aligned} &\langle e^i, y \rangle \geqq 0 \quad \text{for all} \quad i \in I, \\ &\langle y', y \rangle \geqq 0, \\ &\langle y'', y \rangle \geqq 0, \end{aligned} \tag{4.23}$$

coincides with the dual cone D.

III. Find the fundamental system of solutions to the system of linear inequalities (4.23). It must be a collection of vectors whose all linear nonnegative combinations coincide with the solution set of system (4.23). And none of the vectors from the fundamental system can be expressed as N-combination of the other vectors from this system.

Let us specify a certain collection of solutions to the system of linear inequalities (4.23). First of all, note that in space R^m each unit vector e^s for $s \in I \backslash (B \cup C)$ is a solution to (4.23). Next, introduce the vectors

$$e^{ij} = w'_j e^i + w'_i e^j \quad \text{for all } i \in A \text{ and all } j \in B.$$

Their components are nonnegative, and therefore all these vectors satisfy inequalities (4.23) written in the first row. Moreover, they satisfy the inequality $\langle y', y \rangle \geqq 0$ from the second row of (4.23), since

$$\langle y', e^{ij} \rangle = y'_i w'_j + y'_j w'_i = 0 \quad \text{for all } i \in A \text{ and all } j \in B.$$

Finally, the vectors under consideration clearly satisfy the inequality from the third row of (4.23).

In a similar manner, one may check that the vectors

$$\hat{e}^{ik} = w''_k e^i + w''_i e^k \quad \text{for all } i \in A \text{ and all } k \in C$$

satisfy system (4.23).

The vectors $e^{ijk} = w'_j w''_k e^i + w'_i w''_k e^j + w'_j w''_i e^k$ for all $i \in A,\ j \in B,\ k \in C$ also satisfy system (4.23) so that the inequalities in the second and third rows hold as equalities.

Consequently, the collection composed of the vectors e^s for all $s \in I \backslash (B \cup C)$, the vectors e^{ij} for all $i \in A$ and $j \in B$, the vectors $\hat{e}^{ik}$ for all $i \in A$ and $k \in C$, as well as the vectors e^{ijk} for all $i \in A,\ j \in B,\ k \in C$, belongs to the dual cone D. Denote this collection by (*). In addition, as easily observed, none of the vectors from this collection is representable in the form of N-combination of the other vectors. The total number p of all vectors in collection (*) makes up $p = m - |B| - |C| + |A| \cdot |B| + |A| \cdot |C| + |A| \cdot |B| \cdot |C|$.

IV. To verify that the above collection of vectors forms the fundamental system of solutions to the system of linear inequalities (4.23), it remains to show that these inequalities have no other solutions (up to a positive factor) except all N-combinations of the vectors from collection (*). To this effect, together with system (4.23) consider the corresponding system of $(m+2)$ linear equations

$$\begin{aligned} \langle e^i, y \rangle &= 0 \quad \text{for all } i \in I, \\ \langle y', y \rangle &= 0, \\ \langle y'', y \rangle &= 0. \end{aligned} \tag{4.24}$$

By calculating the ranks of the appropriate matrices, we may verify that any subsystem of $(m-1)$ vectors from the collection $e^1, e^2, \ldots, e^m, y', y''$ is linearly independent. Hence, the desired fundamental system of solutions to the system of linear inequalities (4.23) is contained in the one-dimensional nonzero solutions to the subsystems of $(m-1)$ equations from system (4.24).

We will sequentially remove three equations from (4.24) and write the solutions to the resulting subsystems that also satisfy the system of inequalities (4.23). The vectors constructed in this way form the required fundamental system of solutions to the system of inequalities (4.23).

If the last two equations of system (4.24) are among the removed ones, then the unit vectors $e^1, e^2, \ldots, e^m$ give nonzero solutions to the resulting subsystems (up to a positive factor). However, in this collection only the vectors e^s with $s \in I \backslash (B \cup C)$ satisfy the system of inequalities (4.23).

If the last equation of system (4.24) remains in the system and the last but one is eliminated, then the vectors $\hat{e}^{ik}$ for all $i \in A$ and all $k \in C$ are nonzero solutions to the resulting subsystems (up to a positive factor). As established earlier, all these vectors satisfy inequalities (4.23).

Similarly, if the last but one equation remains in (4.24) and the last equation is removed, then the vectors e^{ij} for all $i \in A$ and all $j \in B$ can be taken as solutions of the resulting subsystems. Clearly, these vectors satisfy the system of linear inequalities (4.23).

Finally, imagine that the last two equations remain in the subsystem; in this case, we get the solutions e^{ijk} for all $i \in A,\ j \in B,\ k \in C$.

We have considered all possible triplets of equations that can be eliminated from system (4.24), and there exist no other solutions (up to a positive factor) to the subsystems of $(m - 2)$ equations from system (4.24) that satisfy (4.23). This means that the vector collection (*) forms the fundamental system of solutions to the system of linear inequalities (4.23). Hence, any solution to inequalities (4.23) can be represented in the form of the nonnegative linear combination of this collection. For convenience, denote the vectors in this collection by $a^1, a^2, \ldots, a^p$.

In the case $|A| = 1$ (i.e.,$A = \{i\}$), the arguments are the same and even slightly simpler. Here it is necessary to consider a system of $(m + 1)$ equations obtained from (4.24) by eliminating the equation $\langle e^i, y \rangle = 0$. In this case, we remove sequentially merely two equations from the original system to get the same fundamental system of solutions to the system of linear inequalities (4.23).

V. On the strength of the earlier considerations, the nonzero solution set of the system of linear inequalities (4.23), i.e., the cone D coincides with the set of all N-combinations of the vectors $a^1, a^2, \ldots, a^p$. Therefore, the vector u satisfies the inclusion $u \in D$ if and only if this vector can be represented as some N-combination of the vectors from the above collection.

Owing to this circumstance, the inequality

$$\langle u, y \rangle \geqq 0 \quad \text{for all}\, u \in D, \tag{4.25}$$

where y is an arbitrary fixed nonzero vector, appears equivalent to the inequalities

$$\langle a^i, y \rangle \geq 0, i = 1, 2, \ldots, p. \tag{4.25'}$$

Note that the sign $\geq$ indicates that, at least for one index $i \in \{1,2,\ldots,p\}$, the inequality is strict. Really, let the vector y satisfy inequalities (4.25′). Each vector $u \in D$ can be expressed as some N-combination of the vectors $a^1, a^2, \ldots, a^p$, e.g., $z = \lambda_1 a^1 + \lambda_2 a^2 + \ldots + \lambda_p a^p$. Multiply inequalities (4.25′) by the corresponding nonnegative numbers $\lambda_1, \lambda_2, \ldots, \lambda_p$(that are not zero simultaneously) and perform termwise addition of the resulting inequalities to get (4.25):

$$\left\langle \sum_{i=1}^{p} \lambda_i a^i, y \right\rangle = \langle u, y \rangle \geqq 0.$$

Conversely, inequality (4.25′) follows from (4.25), since $a^i \in D$ for all $i = 1,2,\ldots p$. And inequalities (4.25′) do not hold as equalities all simultaneously. Indeed, assume that inequalities (4.25′) turn out into equalities for a nonzero vector y. Then these inequalities take place for the opposite vector $-y$. Hence, the cone dual to D is not acute. But this dual cone has the form

$$M = \{y \in R^m | \quad \langle u, y \rangle \geqq 0 \quad \text{for all}\, u \in D\} \backslash \{0_m\}, \tag{4.26}$$

since, for polyhedral cones, the dual cone to the dual cone D is the original cone M. We have arrived at a contradiction: the cone M is not acute. This means that, for the nonzero vector y, inequalities (4.25′) do not hold as equalities all simultaneously.

Based on the established equivalence of inequalities (4.25) and (4.25′), from (4.26) we infer that the inclusion $y \in M$ holds if and only if inequalities (4.25′) are true, i.e.,

$$y \in M \Leftrightarrow \langle a^i, y \rangle \geq 0, i = 1,2,\ldots,p.$$

VI. The final stage of the proof is standard; therefore, we omit it. ■

According to Theorem 4.9, for taking into account the information about the DM's preference relation in the form of the two relationships $y' \succ 0_m$ and $y'' \succ 0_m$, reducing the Pareto set based on them, it is necessary to form the new vector criterion g by the above formulas and then construct the Pareto set $P_g(X)$ in terms of this criterion.

Using the proof scheme of Theorem 4.9 and similar considerations, one may further generalize it to the case where the DM is willing to compromise a whole collection of groups of criteria $B_1, B_2, \ldots, B_k$ for gaining in terms of a group of criteria A. In particular, the following result takes place.

Theorem 4.10 *Assume that there is a given finite collection of* $k+1$ *pairwise disjoint subsets of criteria indexes* $A, B_1, B_2, \ldots, B_k \subset I$ *and information from the DM stating that* $y^s \succ 0_m, s = 1,2,\ldots,k$, *where the vectors* y^s *for* $s = 1,2,\ldots,k$ *have the components*

$$y_i^s = w_i^s \quad \textit{for all} \quad i \in A,$$

$$y_j^s = -w_j^s \quad \textit{for all} \quad j \in B_s; \quad y_j^s = 0 \quad \textit{for all} \quad j \in I \backslash (A \cup B_s),$$

where all w_i^s, w_j^s are fixed positive numbers.

Then this information is consistent, and for any set of selectable vectors $C(Y)$ we have inclusions (4.10), *where $\hat{P}(Y) = f(P_g(X))$ gives the subset of feasible vectors corresponding to the set of Pareto optimal alternatives in the multicriteria problem with the initial set X and the new r-dimensional vector criterion g, $r = m - \sum_{s=1}^{k} |B_s| + |A| \cdot \sum_{s=1}^{k} |B_s| + |A| \cdot \prod_{s=1}^{k} |B_s|$, that consists of the components* ($s = 1, 2, \ldots, k$)

$$g_{ij}^s = w_j^s f_i + w_i^s f_j \quad \textit{for all}\, i \in A, j \in B_s,$$

$$g_{ij_1\ldots j_k}^s = w_{j_1}^s \cdot \ldots \cdot w_{j_k}^s f_i + w_i^s \cdot w_{j_2}^s \cdot \ldots \cdot w_{j_k}^s f_{j_1} + \ldots + w_i^s \cdot w_{j_1}^s \cdot \ldots \cdot w_{j_{k-1}}^s f_{j_k}$$
$$\textit{for all}\, i \in A,\ j_1 \in B_1, \ldots, j_k \in B_k,$$

$$g_t = f_t \quad \textit{for all}\, t \in I \backslash \bigcup_{s=1}^{k} B_s.$$

4.3.2 Usage of Information Quanta of Set-Point Type

Now, consider situation where (according to available information about the preferences) the DM is willing to compromise criteria of a third group C for gaining in terms of criteria of two groups A and B. As illustrated by the forthcoming theorem, such information is always consistent and, for taking it into account, we have to construct the "new" vector criterion from the "old" one by replacing all components from the group C with the "new" components calculated by definite formulas.

Theorem 4.11 *Assume that there are three given pairwise disjoint subsets of criteria indexes $A, B, C \subset I$ and information that $y' \succ 0_m$ and $y'' \succ 0_m$, where the vectors y' and y'' have the components*

$$y_i' = w_i' \quad \textit{for all}\ i \in A, \quad y_k' = -w_k' \quad \textit{for all}\ k \in C, \quad y_s' = 0$$
$$\textit{for all}\ s \in I \backslash (A \cup B),$$

$$y_j'' = w_j'' \quad \textit{for all} \quad j \in B, \quad y_k'' = -w_k'' \quad \textit{for all} \quad k \in C, \quad y_s'' = 0$$
$$\textit{for all}\ s \in I \backslash (B \cup C).$$

where all w_i', w_k', w_j'', w_k'' are fixed positive numbers.

Then this information is consistent, and for any set of selectable vectors C(Y) we have inclusions (4.10), *where* $\hat{P}(Y) = f(P_g(X))$ *gives the subset of feasible vectors corresponding to the set of Pareto optimal alternatives in the multicriteria problem with the initial set X and the new* $(m - |C| + |A| \cdot |B| \cdot |C|)$*-dimensional vector criterion g that consists of the components*

$$g_{ijk} = w'_i w''_j f_k + w''_j w'_k f_i + w'_i w''_k f_j \quad \textit{for all } i \in A, j \in B, k \in C, \qquad g_s = f_s \quad \textit{for all } s \in I \backslash C. \tag{4.27}$$

□ The proof consists of four stages as follows.

I. At the beginning, verify the consistency of the available information. By Theorem 4.2, this information is consistent if and only if the system of linear algebraic equations

$$\sum_{i=1}^{m} \lambda_i e^i + \lambda_{m+1} y' + \lambda_{m+2} y'' = 0_m \tag{4.28}$$

has no N-solution $\lambda_1, \ldots, \lambda_{m+2}$. System (4.28) can be rewritten in the expanded form

$$\lambda_i + \lambda_{m+1} w'_i = 0 \quad \text{for all } i \in A,$$

$$\lambda_j + \lambda_{m+2} w''_j = 0 \quad \text{for all } j \in B,$$

$$\lambda_k - \lambda_{m+1} w'_k - \lambda_{m+2} w''_k = 0 \quad \text{for all } k \in C,$$

$$\lambda_l = 0 \quad \text{for all } l \in I \backslash (A \cup B \cup C).$$

Owing to the non-negativeness of $\lambda_i, \lambda_j, \lambda_{m+1}, \lambda_{m+2}$, the equations in the first and second rows imply the equalities $\lambda_i = \lambda_j = \lambda_{m+1} = \lambda_{m+2} = 0$ for all $i \in A$ and $j \in B$. In this case, the other equations yield the equalities $\lambda_k = 0$ for the other indexes $k \in I \backslash (A \cup B)$. Hence, system (4.28) has the trivial (zero) solution as the only nonnegative solution. This establishes the consistency of the existing information in the form of the relationships $y' \succ 0_m$ and $y'' \succ 0_m$.

II. Denote by K the acute convex cone (without the origin) of the cone relation $\succ$, and by M the acute convex cone (without the origin) generated by the vectors $e^1, \ldots, e^m, y', y''$.

Here four cases are possible, namely, $|A| > 1$ and $|B| > 1$; $|A| = 1$ and $|B| > 1$; $|A| > 1$ and $|B| = 1$; and, $|A| = |B| = 1$. In the first case, the generators of the cone M are all the vectors $e^1, e^2, \ldots, e^m, y', y''$, since none of them can be expressed as the N-combination of the other vectors from this collection (due to the acuteness of the cone M). In the second case (i.e., if $A = \{i\}$), the vector e^i can be represented by the linear positive combination of the vector y' and all vectors e^s for $s \in B$. And so, in this case, the generators of the cone M are the vectors $e^1, e^2, \ldots, e^m, y', y''$ without the vector e^i. Similarly, in the third case, the generators of the cone M are the vectors $e^1, e^2, \ldots, e^m, y', y''$ without the vector e^j; in the fourth case, the vectors e^i and e^j are both excluded from the list of generators.

Our analysis begins with the first case. For the polyhedral cone M, introduce the dual cone D (without the origin) of the form

$$D = \{y \in R^m | \langle u, y \rangle \geqq 0 \text{ for all } u \in M\} \backslash \{0_m\}.$$

Since the cone M is generated by the vectors $e^1, e^2, \ldots, e^m, y', y''$, the set of nonzero solutions to the system of linear homogeneous inequalities

$$\begin{aligned} &\langle e^i, y \rangle \geqq 0 \quad \text{for all } i \in I, \\ &\langle y', y \rangle \geqq 0, \\ &\langle y'', y \rangle \geqq 0, \end{aligned} \tag{4.29}$$

coincides with the dual cone D.

III. Find the fundamental system of solutions to the system of linear inequalities (4.29). It must be a minimal collection of vectors whose all N-combinations coincide with the nonzero solution set of system (4.29). And none of the vectors from the fundamental system can be expressed as the N-combination of the other vectors from this system.

Let us specify a certain collection of solutions to the system of linear inequalities (4.29). Clearly, the vectors $e^{ijk} = w_j'' w_k' e^i + w_i' w_k'' e^j + w_j'' w_i' e^k$ for all $i \in A$, $j \in B$, $k \in C$ satisfy system (4.29), and for these vectors the inequalities from the second and third rows hold as equalities.

Consequently, the collection composed of the vectors e^s for all $s \in I \backslash C$ and the vectors e^{ijk} for all $i \in A$, $j \in B$, $k \in C$, belongs to the dual cone D. Denote this collection by (*). In addition, as easily observed, none of the vectors from this collection is representable in the form of the linear nonnegative combination of the other vectors. The total number p of all vectors in collection (*) makes up $p = m - |C| + |A| \cdot |B| \cdot |C|$.

To verify that the above collection of vectors forms the fundamental system of solutions to the system of linear inequalities (4.29), it remains to show that these inequalities have no other solutions (up to a positive factor) except all N-combinations of the vectors from collection (*). To this end, together with system (4.29) consider the corresponding system of $(m+2)$ linear equations

$$\begin{aligned} &\langle e^i, y \rangle = 0 \quad \text{for all } i \in I, \\ &\langle y', y \rangle = 0, \\ &\langle y'', y \rangle = 0. \end{aligned} \tag{4.30}$$

By calculating the ranks of the appropriate matrices, we may check that any subsystem of $(m-1)$ vectors from the collection $e^1, e^2, \ldots, e^m, y', y''$ is linearly independent. Hence, the required fundamental system of solutions to the system of linear inequalities (4.29) is contained in the one-dimensional nonzero solutions to the subsystems of $(m - 1)$ equations from system (4.30).

We will sequentially remove three equations from (4.30) and write the solutions to the resulting subsystems that also satisfy the system of inequalities (4.29). The vectors constructed in this way form the required fundamental system.

If the last two equations of system (4.30) are among the removed ones, then the unit vectors $e^1, e^2, \ldots, e^m$ give nonzero solutions to the resulting subsystems (up to a positive factor). However, in this collection only the vectors e^s with $s \in I \backslash C$ satisfy the system of inequalities (4.29).

If the last equation of system (4.30) remains in the system and the last but one is eliminated, then the vectors not satisfying the system of inequalities (4.29) are nonzero solutions to the resulting "truncated" subsystems (up to a positive factor).

Similarly, if the last but one equation remains in (4.30) and the last equation is removed, then the solutions to the "truncated" subsystems contain no vectors satisfying the system of linear inequalities (4.29).

Finally, imagine that the two last equations remain in the "truncated" subsystem; in this case, we have the solutions e^{ijk} for all $i \in A,\ j \in B,\ k \in C$.

We have considered all possible triplets of equations that can be eliminated from system (4.30), and there exist no other solutions (up to a positive factor) to the subsystems of $(m - 2)$ equations from system (4.30) that satisfy (4.29). There also exist no other one-dimensional solutions to system (4.30) that satisfy (4.29).This means that the vector collection (*) forms the fundamental system of solutions to the system of linear inequalities (4.29). Hence, any nonzero solution to inequalities (4.29) can be represented in the form of N-combination of this collection. For convenience, denote the vectors in this collection by $a^1, a^2, \ldots, a^p$.

In the case $|A| = 1$ (i.e., $A = \{i\}$), the arguments are the same and even slightly simpler. Here it is necessary to consider a system of $(m + 1)$ equations obtained from (4.30) by eliminating the equation $\langle e^i, y \rangle = 0$. In this case, we remove sequentially merely two equations from the original system to get the fundamental system of solutions to the system of linear inequalities (4.29). The other cases are studied by analogy, and we therefore omit them.

IV. The rest parts of the proof are the same as in Theorem 4.9. ∎

Using the proof scheme of Theorem 4.11 and similar considerations, one may further generalize it in the following manner.

Theorem 4.12 *Assume that there are a given finite collection of* $k+1$ *pairwise disjoint subsets of criteria indexes* $A_1, A_2, \ldots, A_k, B \subset I$ *and information stating that* $y^s \succ 0_m,\quad s = 1, 2, \ldots, k$, *where the vectors* y^s *(for* $s = 1, 2, \ldots, k$*) have the components*

$y_i^s = w_i^s$ *for all* $i \in A_s$, $y_j^s = -w_j^s$ *for all* $j \in B$, $y_t^s = 0$ *for all* $t \in I \backslash (A_s \cup B)$ *with positive numbers* w_i^s, w_j^s.

Then this information of set-point type is consistent, and for any set of selectable vectors C(Y) we have the inclusions (4.10), *where* $\hat{P}(Y) = f(P_g(X))$, *and also the new* $(m - |B| + |B| \cdot \prod_{s=1}^{k} |A_s|)$

-dimensional vector criterion g consists of the following components

$g^s_{ji_1 \ldots i_k} = w^s_{i_1} \cdot \ldots \cdot w^s_{i_k} f_j + w^s_j \cdot w^s_{i_2} \cdot \ldots \cdot w^s_{i_k} f_{i_1} + \ldots + w^s_{i_1} \cdot \ldots \cdot w^s_{i_{k-1}} \cdot w^s_j f_{i_k}$ *for all* $i_1 \in A_1, \ldots, i_k \in A_k$, $j \in B$, $s = 1, 2, \ldots, k$,

$g_t = f_t$ *for all* $t \in I \backslash B$.

Chapter 5
Pareto Set Reduction Based on Collections of Information Quanta

Here we further develop the results of the previous chapter on Pareto set reduction using given finite collections of mutually dependent information quanta about the DM's preference relation. In addition, this chapter formulates and solves the problem of taking into account an *arbitrary* finite collection of information quanta. Particularly, we suggest two algorithms to construct a new vector criterion using a given arbitrary finite collection of quanta; the Pareto set in terms of this vector criterion is an upper estimate for the unknown set of selectable alternatives (or vectors). At the end of this chapter, we describe an algorithm of taking into account an arbitrary collection of information quanta for the set of feasible vectors consisting of a finite number of elements.

5.1 Closed Collections of Information Quanta

5.1.1 Closed Collection of Two Information Quanta and Its Consistency

First, introduce the notion of a mutually dependent closed collection of two information quanta about the DM's preference relation.

Definition 5.1 Let $m \geqq 3$ and consider two groups A and B that consist of r and t criteria, respectively, where $r+t \leqq m$ and $A \cap B = \varnothing$. Without loss of generality, it is possible to renumber the criteria so that the group A contains the criteria of indexes $\{1, 2, \ldots, r\}$, whereas the group B the criteria of indexes $\{r+1, \ldots, r+t\}$. Assume that the group of criteria A is more important than the group B with the collections of positive parameters w_i for all $i \in A$ and w_j for all $j \in B$. In addition, assume that the group of criteria $B = \{r+1, \ldots, r+t\}$ is more important than the group of criteria $A = \{1, 2, \ldots, r\}$ with the collections of positive

© Springer International Publishing AG 2018

V.D. Noghin, *Reduction of the Pareto Set*, Studies in Systems, Decision and Control 126, https://doi.org/10.1007/978-3-319-67873-3_5

parameters γ_j for all $j \in B$ and γ_i for all $i \in A$. In this case, we say that there is a given *closed collection of two information quanta* (*A*, *B*).

According to the simplified definition of an information quantum about the DM's preference relation, the existence of such a closed collection of two information quanta (*A*, *B*) is equivalent to the specification of two vectors

$y^1 = (w_1, \ldots, w_r, -w_{r+1}, \ldots, -w_{r+t}, 0, \ldots, 0)$ and
$y^2 = (-\gamma_1, \ldots, -\gamma_r, \gamma_{r+1}, \ldots, \gamma_{r+t}, 0, \ldots, 0)$
that satisfy the relationships $y^1 \succ 0_m$ and $y^2 \succ 0_m$.

Lemma 5.1 *A closed collection of two information quanta* (*A*, *B*) *is consistent if and only if there exist criteria indexes* $p \in A$ *and* $l \in B$ *such that*

$$\frac{w_p}{\gamma_p} > \frac{w_l}{\gamma_l}. \tag{5.1}$$

□ This result will be proved by establishing that a closed collection of information (*A*, *B*) is inconsistent if and only if

$$\frac{w_i}{\gamma_i} \leqq \frac{w_j}{\gamma_j} \quad \text{for all } i \in A \text{ and } j \in B. \tag{5.2}$$

Necessity. If a closed collection of information is inconsistent, then by the algebraic criterion of consistency the system of homogeneous linear equations

$$\lambda_1 e^1 + \ldots + \lambda_m e^m + \lambda_{m+1} y^1 + \lambda_{m+2} y^2 = 0_m \tag{5.3}$$

has at least one *N*-solution. Hence, for some nonnegative numbers $\lambda_1^*, \lambda_2^*, \ldots, \lambda_{m+2}^*$ that are not zero all simultaneously, we have

$$\begin{gathered} \lambda_{m+1}^* w_1 - \lambda_{m+2}^* \gamma_1 \leqq 0, \ldots, \lambda_{m+1}^* w_r - \lambda_{m+2}^* \gamma_r \\ \leqq 0, -\lambda_{m+1}^* w_{r+1} + \lambda_{m+2}^* \gamma_{r+1} \leqq 0, \ldots, -\lambda_{m+1}^* w_{r+t} + \lambda_{m+2}^* \gamma_{r+t} \leqq 0. \end{gathered} \tag{5.4}$$

From equality (5.3) it follows that $\lambda_{r+t+1}^* = \lambda_{r+t+2}^* = \ldots = \lambda_m^* = 0$, $(\lambda_{m+1}^*, \lambda_{m+2}^*) \geq 0_2$ and

$$\lambda_{m+1}^{*}w_1 \leqq \lambda_{m+2}^{*}\gamma_1, \ldots, \lambda_{m+1}^{*}w_r \leqq \lambda_{m+2}^{*}\gamma_r, -\lambda_{m+1}^{*}w_{r+1} \leqq \lambda_{m+2}^{*}\gamma_{r+1}, \ldots, -\lambda_{m+1}^{*}w_{r+t} \leqq \lambda_{m+2}^{*}\gamma_{r+t}.$$

Without loss of generality we may assume $\lambda_{m+2}^{*} > 0$. After trivial transformations, we obtain

$$\frac{\lambda_{m+1}^{*}}{\lambda_{m+2}^{*}} \leqq \frac{\gamma_1}{w_1}, \ldots, \frac{\lambda_{m+1}^{*}}{\lambda_{m+2}^{*}} \leqq \frac{\gamma_r}{w_r}, \frac{\lambda_{m+1}^{*}}{\lambda_{m+2}^{*}} \geqq \frac{\gamma_{r+1}}{w_{r+1}}, \ldots, \frac{\lambda_{m+1}^{*}}{\lambda_{m+2}^{*}} \geqq \frac{\gamma_{r+t}}{w_{r+t}}. \tag{5.5}$$

By eliminating the quantities $\lambda_{m+1}^{*}, \lambda_{m+2}^{*}$, system (5.5) can be rewritten as

$$\frac{\gamma_j}{w_j} \leqq \frac{\gamma_i}{w_i} \quad \text{for all } i \in A \text{ and } j \in B.$$

Obviously, this system of inequalities is equivalent to (5.2).

Sufficiency. Let inequalities (5.2) hold. In this case, there exist positive numbers $\lambda_{m+1}^{*}, \lambda_{m+2}^{*}$ satisfying (5.5). By repeating the same steps as in the necessity part (now, inversely), we arrive at formula (5.4). Therefore, the system of linear Eq. (5.3) has the *N*-solution $\lambda_1^{*}, \lambda_2^{*}, \ldots, \lambda_{m+2}^{*}$. Subsequently, the closed collection of the two information quanta (*A*, *B*) is inconsistent. ■

Introduce the following sets of criteria indexes:

$$P = \{p \in A|\ \text{there exists}\, j \in B \ \text{ such that } \ \frac{w_p}{\gamma_p} > \frac{w_j}{\gamma_j}\},$$

$$L = \{l \in B|\ \text{there exists } i \in A \text{ such that } \ \frac{w_i}{\gamma_i} > \frac{w_l}{\gamma_l}\},$$

$$P_l = \{p \in P|\ \text{for fixed } l \in L, \frac{w_p}{\gamma_p} > \frac{w_l}{\gamma_l}\},$$

$$L_p = \{l \in L|\ \text{for fixed } p \in P, \frac{w_p}{\gamma_p} > \frac{w_l}{\gamma_l}\},$$

$$\bar{P}_l = \{i \in A|\ \text{for fixed } l \in L, \frac{w_i}{\gamma_i} \leqq \frac{w_l}{\gamma_l}\},$$

$$\bar{L}_p = \{j \in B|\ \text{for fixed } p \in P, \frac{w_p}{\gamma_p} \leqq \frac{w_j}{\gamma_j}\}.$$

According to the above notation, for each $l \in L$ we have the relationships $P_l \cup \bar{P}_l = A$ and $P_l \cap \bar{P}_l = \varnothing$; for each $p \in P$, the relatissonships $L_p \cup \bar{L}_p = B$, $L_p \cap \bar{L}_p = \varnothing$, and $|P| \leqq r, |L| \leqq t$. Here the overline symbol indicates the complement of an appropriate set to the criteria index set $\{1, 2, \ldots, m\}$.

Remark 5.1 Lemma 5.1 says that a closed collection of two information quanta (*A*, *B*) is consistent if and only if $P \neq \varnothing$ and $L \neq \varnothing$.

5.1.2 Reduction of the Pareto Set Using Closed Collection

The following result is the case.

Theorem 5.1 *Assume that there is a given consistent (i.e., $P \neq \emptyset$ and $L \neq \emptyset$) closed collection of two information quanta (A, B). Then for any set of selectable vectors $C(Y)$ we have the inclusions*

$$C(Y) \subset \hat{P}(Y) \subset P(Y), \tag{5.6}$$

where $\hat{P}(Y) = f(P_g(X))$ and the vector criterion g of the dimension

$$m - (|A| + |B|) + \sum_{p \in P} |L_p| + \sum_{l \in L} |P_l| + \sum_{l \in L} |P_l||\bar{P}_l| + \sum_{p \in P} |L_p||\bar{L}_p| \tag{5.7}$$

has the components

$$g_{pl} = w_l f_p + w_p f_l \quad \text{for all } p \in P \text{ and } l \in L \text{ satisfying}(5.1),$$
$$g_{lp} = \gamma_l f_p + \gamma_p f_l \quad \text{for all } p \in P \text{ and } l \in L \text{ satisfying}(5.1),$$

$$\begin{aligned} g_{pli} &= (\gamma_i w_l - \gamma_l w_i) f_p + (\gamma_l w_p - \gamma_p w_l) f_i \\ &\quad + (\gamma_i w_p - \gamma_p w_i) f_l \quad \text{for all } l \in L, p \in P_l, \text{ and } i \in \bar{P}_l, \\ g_{plj} &= (\gamma_l w_j - \gamma_j w_l) f_p + (\gamma_l w_p - \gamma_p w_l) f_j \\ &\quad + (\gamma_p w_j - \gamma_j w_p) f_l \quad \text{for all } p \in P, l \in L_p, \text{ and } j \in \bar{L}_p, \\ g_s &= f_s \quad \text{for all } s \in I \backslash \{A \cup B\}. \end{aligned} \tag{5.8}$$

I. As stated earlier, the specification of a closed collection of two information quanta (A, B) means that the cone K of the cone preference relation $\succ$ includs the m-dimensional vectors

$$y^1 = (w_1, w_2, \ldots, w_r, -w_{r+1}, -w_{r+2}, \ldots, -w_{r+t}, 0, \ldots, 0),$$
$$y^2 = (-\gamma_1, -\gamma_2, \ldots, -\gamma_r, \gamma_{r+1}, \gamma_{r+2}, \ldots, \gamma_{r+t}, 0, \ldots, 0).$$

Moreover, owing to the Pareto axiom, we have $R^m_+ \subset K$.

Introduce the cone M (without the origin) generated by the collection of the vectors $e^1, e^2, \ldots, e^m, y^1, y^2$. Due to Theorem 4.1 the cone M is acute and convex.

I. Only one of the following four cases is possible:

(1) $|P| = 1, |L| = 1$, (2) $|P| = 1, |L| > 1$, (3) $|P| > 1, |L| = 1$, (4) $|P| > 1, |L| > 1$.

We begin with the first case where the sets P and L are singletons, i.e., $P = \{p\}, L = \{l\}$. Then

$$\begin{aligned} \tfrac{w_p}{\gamma_p} > \tfrac{w_l}{\gamma_l}, \tfrac{w_i}{\gamma_i} \leqq \tfrac{w_j}{\gamma_j} \quad & \text{for all } i \in A\backslash\{p\}, j \in B, \\ \tfrac{w_i}{\gamma_i} \leqq \tfrac{w_j}{\gamma_j} \quad & \text{for all } j \in B\backslash\{l\}, i \in A. \end{aligned} \tag{5.9}$$

Show that the generators of the cone M are the vectors y^1, y^2 and the unit vectors e^s for all $s \in I\backslash\{P \cup L\}$. Supposing the opposite, represent, e.g., the vector e^s for $s \in (A \cup B)\backslash\{P \cup L\}$ in the form

$$e^s = \sum_{\substack{k=1, \\ k \neq s}}^{m} \lambda_k e^k + \mu_1 y^1 + \mu_2 y^2, \tag{5.10}$$

where the coefficients $\lambda_k, 1, 2, \ldots, m; k \neq s, y^1, y^2$ are nonnegative. Consider components p and l of this vector equality, which satisfy (5.1):

$$-\lambda_p = \mu_1 w_p - \mu_2 \gamma_p, \quad -\lambda_l = -\mu_1 w_l + \mu_2 \gamma_l,$$

or

$$\mu_1 w_p - \mu_2 \gamma_p \leqq 0, \quad -\mu_1 w_l + \mu_2 \gamma_l \leqq 0.$$

The last inequalities yield

$$\frac{w_l}{\gamma_l} \geqq \frac{w_p}{\gamma_p},$$

which contradicts (5.1).

If the vector e^s is such that $s \in I\backslash(A \cup B)$, then contradiction directly follows from (5.10), notably, $1 = \sum_{\substack{k=1 \\ k \neq s}}^{m} \lambda_k \cdot 0 + \mu_1 \cdot 0 + \mu_2 \cdot 0$.

Now, assume that the vector y^1 can be expressed as

$$y^1 = \sum_{k=1}^{m} \lambda_k e^k + \mu y^2, \mu \geqq 0.$$

In this case, for each component $j \in B$ we have the contradictory relationship $0 > -w_j = \lambda_j + \mu \gamma_j \geqq 0$.

In a similar manner, the vector y^2 is not representable as N-combination of the others vectors.

Demonstrate that the vector e^p can be expressed as N-combination of the vectors y^1, y^2 and e^s, where $s \in I \backslash \{p, l\}$. A necessary and sufficient condition is that there exist nonnegative numbers μ_1, μ_2, λ_s that are not zero all simultaneously and satisfy the equality

$$e^p = \sum_{s \in I \backslash (p,l)} \lambda_s e^s + \mu_1 y^1 + \mu_2 y^2,$$

or (in the coordinatewise form)

$$\begin{gathered} 1 = \mu_1 w_p - \mu_2 \gamma_p, \\ 0 = \lambda_i + \mu_1 w_i - \mu_2 \gamma_i \quad \text{for all } i \in A \backslash \{p\}, \\ 0 = \lambda_j - \mu_1 w_j + \mu_2 \gamma_j \quad \text{for all } j \in B \backslash \{l\}, \\ 0 = -\mu_1 w_l + \mu_2 \gamma_l, \\ 0 = \lambda_s \quad \text{for all } s \in I \backslash (A \cup B). \end{gathered}$$

This system of equations has the unique solution

$$\mu_1 = \frac{\gamma_l}{\gamma_l w_p - \gamma_p w_l}, \ \mu_2 = \frac{w_l}{\gamma_l w_p - \gamma_p w_l}, \ \lambda_i = \frac{\gamma_i w_l - \gamma_l w_i}{\gamma_l w_p - \gamma_p w_l} \quad \text{for all } i \in A \backslash \{p\},$$
$$\lambda_j = \frac{\gamma_l w_j - \gamma_j w_l}{\gamma_l w_p - \gamma_p w_l} \quad \text{for all } j \in B \backslash \{l\}, \ \lambda_s = 0 \quad \text{for all } s \in I \backslash (A \cup B).$$

The above solution is N-solution if and only if

$$\frac{w_p}{\gamma_p} > \frac{w_l}{\gamma_l}, \ \frac{w_i}{\gamma_i} \leqq \frac{w_l}{\gamma_l} \quad \text{for all } i \in A \backslash \{p\}, \ \frac{w_j}{\gamma_j} \geqq \frac{w_l}{\gamma_l} \quad \text{for all } j \in B \backslash \{l\}.$$

By virtue of (5.9), all these inequalities hold. Therefore, the vector e^p can be expressed as N-combination of the vectors y^1, y^2 and e^s for all $s \in I \backslash \{p, l\}$.

Using the same procedure, we can verify that the vector e^l is representable as N-combination of the vectors y^1, y^2 and e^s for all $s \in I \backslash \{p, l\}$.

Consider the second case when $P = \{p\}$, $|L| > 1$. Then the following inequalities take place:

$$\frac{w_p}{\gamma_p} > \frac{w_l}{\gamma_l} \quad \text{for all } l \in L, \ \frac{w_i}{\gamma_i} \leqq \frac{w_j}{\gamma_j} \quad \text{for all } i \in A \backslash \{p\} \text{ and } j \in B. \tag{5.11}$$

We will show that the cone M is generated by the vectors y^1, y^2 together with the unit vectors e^s for all $s \in I \backslash P$.

Just like in the first case, verify that each of the vectors y^1, y^2, e^s for all $s \in I \backslash \{P \cup L\}$ is not representable as N-combination of the other vectors.

Fix a number $\bar{l} \in L$ such that $\frac{w_{\bar{l}}}{\gamma_{\bar{l}}} \geqq \frac{w_l}{\gamma_l}$ for all $l \in L$. Show that the vector e^p can be expressed as N-combination of the vectors y^1, y^2, and e^s, $s \in I \backslash \{p, \bar{l}\}$, i.e.,

$$e^p = \sum_{s \in I \setminus \{p, \bar{l}\}} \lambda_s e^s + \mu_1 y^1 + \mu_2 y^2.$$

In this case, it is easy to verify that the numbers μ_1, μ_2, and λ_s (for all $s \in I \setminus \{p, \bar{l}\}$) form N-solution if and only if

$$\frac{w_p}{\gamma_p} > \frac{w_{\bar{l}}}{\gamma_{\bar{l}}}, \frac{w_i}{\gamma_i} \leqq \frac{w_{\bar{l}}}{\gamma_{\bar{l}}} \quad \text{for all } i \in A \setminus \{p\}, \frac{w_j}{\gamma_j} \geqq \frac{w_{\bar{l}}}{\gamma_{\bar{l}}} \quad \text{for all } j \in B \setminus \{\bar{l}\}.$$

Due to (5.11) in this case all above inequalities hold. Therefore, the vector e^p can be expressed as N-combination of the vectors y^1, y^2, and e^s, $s \in I \setminus \{p, \bar{l}\}$.

In contrast to the first case, none of the vectors e^l (for $l \in L$) can be represented as N-combination of the vectors y^1, y^2, and e^s for all $s \in I \setminus \{p, l\}$. Conjecture the opposite. Without loss of generality, for $\bar{l} \in L$ let the following representation be true:

$$e^{\bar{l}} = \sum_{s \in I \setminus \{p, \bar{l}\}} \lambda_s e^s + \mu_1 y^1 + \mu_2 y^2.$$

This system has the solution

$$\mu_1 = \frac{\gamma_p}{\gamma_{\bar{l}} w_p - \gamma_p w_{\bar{l}}}, \quad \mu_2 = \frac{w_p}{\gamma_{\bar{l}} w_p - \gamma_p w_{\bar{l}}}, \quad \lambda_i = \frac{\gamma_i w_p - \gamma_p w_i}{\gamma_{\bar{l}} w_p - \gamma_p w_{\bar{l}}} \quad \text{for all } i \in A \setminus \{p\},$$
$$\lambda_j = \frac{\gamma_p w_j - \gamma_j w_p}{\gamma_{\bar{l}} w_p - \gamma_p w_{\bar{l}}} \quad \text{for all } j \in B \setminus \{\bar{l}\}, \quad \lambda_s = 0 \quad \text{for all } s \in I \setminus (A \cup B).$$

It is nonzero nonnegative if and only if

$$\frac{w_p}{\gamma_p} > \frac{w_{\bar{l}}}{\gamma_{\bar{l}}}, \quad \frac{w_p}{\gamma_p} \geqq \frac{w_i}{\gamma_i} \quad \text{for all } i \in A \setminus \{p\}, \quad \frac{w_j}{\gamma_j} \geqq \frac{w_p}{\gamma_p} \quad \text{for all } j \in B \setminus \{\bar{l}\}.$$

However, then the inequalities $\frac{w_j}{\gamma_j} \geqq \frac{w_p}{\gamma_p}$ fail for $j \in L \setminus \{\bar{l}\}$. This contradiction means that, in the second case, the vectors y^1, y^2, e^s for all $s \in I \setminus P$, are the generators of the cone M.

By analogy, we may consider the third case $|P| > 1$, $|L| = 1$, establishing that the the cone M is generated by the vectors y^1, y^2 and the unit vectors e^s for all $s \in I \setminus L$.

The fourth case $|P| > 1$, $|L| > 1$ actually generalizes the second and third ones, and here each of the vectors $e^1, e^2, \ldots, e^m, y^1, y^2$ is a generator of the cone M.

III. Introduce a cone C (without the origin) that is dual to the cone M, i.e.,

$$C = \{x \in R^m | \langle x, y \rangle \geqq 0 \quad \text{for all } y \in M\} \setminus \{0_m\}.$$

Based on the duality theory of convex analysis, the generators of a polyhedral cone C are the inner normals to the $(m-1)$-dimensional facets of the cone M; conversely, the generators of a cone M are the inner normals to the $(m-1)$-dimensional facets of the cone C.

Consider the first case and prove that the cone C coincides with the set of all nonzero solutions to the system of linear homogtnious inequalities

$$\begin{gathered} \langle e^k, y\rangle \geqq 0 \quad \text{for all } k = I\backslash\{p, l\}, \\ \langle y^1, y\rangle \geqq 0, \\ \langle y^2, y\rangle \geqq 0. \end{gathered} \tag{5.12}$$

To this end, find the fundamental system of solutions to the system of inequalities (5.12) up to a positive factor.

Obviously, the unit vectors e^s for all $s \in I\backslash\{A \cup B\}$ are solutions of system (5.12). The number of such vectors makes up $m - (|A| + |B|)$. In addition, another solution is the vector $y = d_{pl} = w_l e^p + w_p e^l$. As direct substitution shows, the vector d_{pl} satisfies the first $(m-1)$ inequalities of system (5.12). Since the closed collection of two information quanta (A, B) is consistent, this vector also obeys the last inequality of system (5.12). Really, direct substitution with inequality (5.1) yield the required result

$$\langle y^2, d_{pl}\rangle = -\gamma_p w_l + \gamma_l w_p \geqq 0.$$

A solution of the system of inequalities (5.12) is also the vector $y = h_{lp} = \gamma_l e^p + \gamma_p e^l$; this fact can be verified just like the case involving the vector d_{pl}.

Finally, other solutions of system (5.12) are the vectors

$$\begin{aligned} y = q_i &= (\gamma_i w_l - \gamma_l w_i)e^p + (\gamma_l w_p - \gamma_p w_l)e^i \\ &\quad + (\gamma_i w_p - \gamma_p w_i)e^l \quad \text{for all } i \in \bar{P}_l, \\ y = h_j &= (\gamma_l w_j - \gamma_j w_l)e^p + (\gamma_l w_p - \gamma_p w_l)e^j \\ &\quad + (\gamma_p w_j - \gamma_j w_p)e^l \quad \text{for all } j \in \bar{L}_p. \end{aligned}$$

In this case, $\bar{P}_l = A\backslash\{p\}$ and $\bar{L}_p = B\backslash\{l\}$. Direct substitution shows that $y = q_i$ and $y = h_j$ satisfy the last two inequalities of system (5.12). Clearly, owing to (5.9), the collection of vectors q_i, h_j is a solution of the first $(m-2)$ inequalities of system (5.12).

Thereby, we have found the following solutions of the system of linear inequalities (5.12):

$$
\begin{aligned}
& e^s \quad \text{for all } s \in I \backslash \{A \cup B\}, \\
& w_l e^p + w_p e^l, \quad \gamma_l e^p + \gamma_p e^l, \\
& \qquad q_i = (\gamma_i w_l - \gamma_l w_i) e^p + (\gamma_l w_p - \gamma_p w_l) e^i \\
& \qquad\qquad + (\gamma_i w_p - \gamma_p w_i) e^l \quad \text{for all } i \in \bar{P}_l, \\
& \qquad h_j = (\gamma_l w_j - \gamma_j w_l) e^p + (\gamma_l w_p - \gamma_p w_l) e^j \\
& \qquad\qquad + (\gamma_p w_j - \gamma_j w_p) e^l \quad \text{for all } j \in \bar{L}_p,
\end{aligned}
\tag{5.13}
$$

The total number of these solutions is

$$
m - (|A| + |B|) + 2 + |\bar{P}_l| + |\bar{L}_p|. \tag{5.14}
$$

Now, we have to verify that system (5.12) admits no other nonzero solutions except the ones representable by all N-combinations of vectors (5.13). Consider the corresponding system of equations

$$
\begin{aligned}
\langle e^k, y \rangle = 0 \quad & \text{for all } k \in I \backslash \{p, l\}, \\
\langle y^1, y \rangle &= 0, \\
\langle y^2, y \rangle &= 0.
\end{aligned}
\tag{5.15}
$$

Any subcollection composed of $(m-1)$ vectors from the collection y^1, y^2, and e^k for all $k \in I \backslash \{p, l\}$ is linearly independent. Therefore, to find the fundamental system of solutions to the system of inequalities (5.12), it suffices to look through the nonzero solutions to all possible subsystems of system (5.15) that consist of $(m-1)$ equations. And then, among the obtained solutions, it is necessary to choose the ones satisfying (5.12).

Remove the last equation from (5.15). Then the resulting "truncated" subsystem has the solution $w_l e^p + w_p e^l$. If we eliminate equation $(m-1)$, the solution is $\gamma_l e^p + \gamma_p e^l$.

Next, let us remove sequentially one of the first $(m-2)$ equations from system (5.15). This yields the solution set e^s for all $s \in I \backslash \{A \cup B\}$, and

$$
\begin{aligned}
q_i &= (\gamma_i w_l - \gamma_l w_i) e^p + (\gamma_l w_p - \gamma_p w_l) e^i \\
&\qquad + (\gamma_i w_p - \gamma_p w_i) e^l \quad \text{for all } i \in \bar{P}_l, \\
h_j &= (\gamma_l w_j - \gamma_j w_l) e^p + (\gamma_l w_p - \gamma_p w_l) e^j \\
&\qquad + (\gamma_p w_j - \gamma_j w_p) e^l \quad \text{for all } j \in \bar{L}_p.
\end{aligned}
$$

Consider the second case when $|P| = 1$, $|L| > 1$. Here system (5.12) is replaced by the system of inequalities

$$\begin{aligned} \langle e^k, y \rangle &\geqq 0 \quad \text{for all } k \in I \backslash \{p\}, \\ \langle y^1, y \rangle &\geqq 0, \\ \langle y^2, y \rangle &\geqq 0. \end{aligned} \tag{5.16}$$

Just like before, it is possible to show that the fundamental system of solutions to inequalities (5.16) consists of the vectors

$$\begin{aligned} e^s &\quad \text{for all } s \in I \backslash \{A \cup B\}, \\ w_l e^p + w_p e^l &\quad \text{for all } l \in L_p, \\ \gamma_l e^p + \gamma_p e^l &\quad \text{for all } l \in L_p, \\ q_i = (\gamma_i w_l - \gamma_l w_i) e^p &+ (\gamma_l w_p - \gamma_p w_l) e^i \\ + (\gamma_i w_p - \gamma_p w_i) e^l &\quad \text{for all } l \in L_p, i \in \bar{P}_l, \\ h_j = (\gamma_l w_j - \gamma_j w_l) e^p &+ (\gamma_l w_p - \gamma_p w_l) e^j \\ + (\gamma_p w_j - \gamma_j w_p) e^l l \in L &\quad \text{for all } l \in L_p, j \in \bar{L}_p. \end{aligned} \tag{5.17}$$

The total number of these vectors makes up

$$m - (|A| + |B|) + 2|L_p| + |L_p||\bar{P}_l| + |L_p||\bar{L}_p|. \tag{5.18}$$

In the third case, the fundamental system of solutions to the inequalities

$$\begin{aligned} \langle e^k, y \rangle &\geqq 0 \quad \text{for all } k = I \backslash \{l\}, \\ \langle y^1, y \rangle &\geqq 0, \\ \langle y^2, y \rangle &\geqq 0, \end{aligned}$$

is the collection of vectors

$$\begin{aligned} e^s &\quad \text{for all } s \in I \backslash \{A \cup B\}, \\ w_l e^p + w_p e^l &\quad \text{for all } p \in P_l, \\ \gamma_l e^p + \gamma_p e^l &\quad \text{for all } p \in P_l, \\ q_i = (\gamma_i w_l - \gamma_l w_i) e^p &+ (\gamma_l w_p - \gamma_p w_l) e^i \\ + (\gamma_i w_p - \gamma_p w_i) e^l &\quad \text{for all } p \in P_l, i \in \bar{P}_l, \\ h_j = (\gamma_l w_j - \gamma_j w_l) e^p &+ (\gamma_l w_p - \gamma_p w_l) e^j \\ + (\gamma_p w_j - \gamma_j w_p) e^l &\quad \text{for all } p \in P_l, j \in \bar{L}_p. \end{aligned} \tag{5.19}$$

Their total number is

$$m - (|A| + |B|) + 2|L_p| + |L_p||\bar{P}_l| + |L_p||\bar{L}_p|. \tag{5.20}$$

In the fourth case, the fundamental system of solutions to the inequalities

$$\begin{aligned} \langle e^k, y \rangle &\geqq 0 \quad \text{for all } k \in I, \\ \langle y^1, y \rangle &\geqq 0, \\ \langle y^2, y \rangle &\geqq 0, \end{aligned}$$

is the collection of vectors

$$\begin{aligned} e^s & \quad \text{for all } s \in I \backslash \{A \cup B\}, \\ w_l e^p + w_p e^l & \quad \text{for all } p \in P \text{ and } l \in L_p, \\ \gamma_l e^p + \gamma_p e^l & \quad \text{for all } l \in L \text{ and } p \in P_l, \\ q_i = (\gamma_i w_l - \gamma_l w_i) e^p &+ (\gamma_l w_p - \gamma_p w_l) e^i \\ + (\gamma_i w_p - \gamma_p w_i) e^l & \quad \text{for all } l \in L, p \in P_l, i \in \bar{P}_l, \\ h_j = (\gamma_l w_j - \gamma_j w_l) e^p &+ (\gamma_l w_p - \gamma_p w_l) e^j \\ + (\gamma_p w_j - \gamma_j w_p) e^l & \quad \text{for all } p \in P, l \in L_p, j \in \bar{L}_p. \end{aligned} \tag{5.21}$$

Their total number makes up

$$m - (|A| + |B|) + \sum_{p \in P} |L_p| + \sum_{l \in L} |P_l| + \sum_{l \in L} |P_l||\bar{P}_l| + \sum_{p \in P} |L_p||\bar{L}_p|.$$

Obviously, the fundamental system of solutions to (5.21) coincides with (5.13), (5.17), and (5.19) in the first, second, and third cases, respectively, while formulas (5.14), (5.18), and (5.20) for the total number of vectors are special cases of formula (5.7). In particular, in the first case when $P = \{p\}$ and $L = \{l\}$, we have $|L_p| = 1$, $|P_l| = 1$, $|\bar{P}_l| = |A| - 1$, and $|\bar{L}_p| = |B| - 1$. With these equalities, formula (5.14) coincides with (5.7).

Hence, further considerations can be confined to the system of solutions (5.21) only.

V. For convenience, denote by $a^1, a^2, \ldots, a^q$ the vectors of the fundamental system (5.21). Then any vector $z \in C$ can be represented as the N-combination:

$$z = \sum_{i=1}^{q} \lambda_i a^i.$$

We will demonstrate that the cone M is the solution set of definite system of linear inequalities with at least one strict inequality, i.e.,

$$M = \{y \in R^m | \langle a^i, y \rangle \geq 0, \, i = 1, \ldots, q\}. \tag{5.22}$$

Really, let y be an arbitrary nozero vector from M. Then for any $z \in C$ we have the inequality $\langle z, y \rangle \geqq 0$. But $a^i \in C$, $i = 1, \ldots, q$, and therefore

$$\langle a^i, y \rangle \geqq 0\,,\ i = 1, \ldots, q. \tag{5.23}$$

Conjecture that all inequalities in (5.23) become equalities; then the resulting system of equalities has the opposite vector $-y$ as its solution. And we arrive at a contradiction: the cone M is not acute. And so, $M \subset \{y \in R^m | \langle a^i, y \rangle \geq 0\,,\ i = 1, \ldots, q\}$.

Now, verify the reverse inclusion. To this end, choose an arbitrary vector $z \in C$, which is the N-combination $z = \sum_{i=1}^{q} \lambda_i a^i$. By multiplying each of inequalities (5.23) by an appropriate number λ_i and performing their termwise addition, we obtain the inequality $\left\langle \sum_{i=1}^{q} \lambda_i a^i, y \right\rangle \geqq 0$, i.e., $\langle z, y \rangle \geqq 0$, for any $z \in C$. This means that the vector y belongs to the cone that is dual to the cone C. This dual cone is equal to the cone M [57]. Thus, we obtain $y \in M$.

Consequently, the cone M can be rewritten as (5.22).

V. The inclusions $R^m_+ \subset M \subset K$ obviously imply the inclusions

$$\text{Ndom}(Y) \subset \hat{P}(Y) \subset P(Y), \tag{5.24}$$

where

$\hat{P}(Y) = \{y^* \in Y |$ there exists no $y \in Y$ such that $y - y^* \in M\}$.

Consider two arbitrary alternatives $x, x' \in X$ and the corresponding vectors $y = f(x)$, $y' = f(x')$, assuming that $f(x) \neq f(x')$. Due to equality (5.22), the inclusion $y - y' \in M$ takes place if and only if

$$\langle a^i, f(x) - f(x') \rangle \geq 0\,,\ i = 1, \ldots, q,$$

or, equivalently,

$$\langle a^i, f(x) \rangle \geq \langle a^i, f(x') \rangle\,,\ i = 1, \ldots, q,$$

with at least one strict inequality. Recall that the notation a^i is used for vectors (5.21). Hence,

$$\langle w_j e^i + w_i e^j, f(x) \rangle \geqq \langle w_j e^i + w_i e^j, f(x') \rangle \quad \text{for all}\, p \in P\,,\ l \in L_p,$$
$$\langle \gamma_j e^i + \gamma_i e^j, f(x) \rangle \geqq \langle \gamma_j e^i + \gamma_i e^j, f(x') \rangle \quad \text{for all}\, p \in P\,,\ l \in L_p,$$

$$\langle (w_l \gamma_i - w_i \gamma_l) e^p + (w_p \gamma_l - w_l \gamma_p) e^i + (w_p \gamma_i - \gamma_p w_i) e^l, f(x) \rangle$$
$$\geqq \langle (w_l \gamma_i - w_i \gamma_l) e^p + (w_p \gamma_l - w_l \gamma_p) e^i + (w_p \gamma_i - \gamma_p w_i) e^l, f(x') \rangle \quad \text{for all } l \in L, p \in P_l,\ i \in \bar{P}_l,$$

$$\langle (w_j\gamma_l - w_l\gamma_j)e^p + (w_p\gamma_l - w_l\gamma_p)e^j + (w_p\gamma_j - \gamma_p w_j)e^l, f(x)\rangle$$
$$\geqq \langle (w_j\gamma_l - w_l\gamma_j)e^p + (w_p\gamma_l - w_l\gamma_p)e^j + (w_p\gamma_j - \gamma_p w_j)e^l, f(x')\rangle \quad \text{for all } p \in P, l \in L_p, j \in \bar{L}_p,$$

$$\langle e^s, f(x)\rangle \geqq \langle e^s, f(x'')\rangle \quad \text{for all } s \in I\backslash\{A \cup B\},$$

or $g(x) \geq g(x')$, where the vector function g described by (5.8) has dimension (5.7). Consequently, $\hat{P}(Y) = f(P_g(X))$.

Finally, inclusions (5.24) and $C(Y) \subset \text{Ndom}(Y)$ yield (5.6). ■

As a matter of fact, there are two extreme cases when the consistency condition from Lemma 5.1 holds. In the first case, the inequality $\frac{w_i}{\gamma_i} > \frac{w_j}{\gamma_j}$ takes place for all $i \in A$ and $j \in B$. Then $P = A$, $L = B$, $\bar{P}_l = \varnothing$, and $P_l = A$ for each $l \in L$, while $\bar{L}_p = \varnothing$ and $L_p = B$ for each $p \in P$. Accordingly, Theorem 5.1 acquires the following form.

Theorem 5.2 *Assume that there is a given consistent closed collection of two information quanta (A, B) and inequalities* (5.1) *hold for all indexes* $i \in A$ *and* $j \in B$. *Then for any set of selectable vectors C(Y) we have inclusions* (5.6), *where* $\hat{P}(Y) = f(P_g(X))$ *and the vector criterion g of dimension* $m - (|A| + |B|) + 2|A||B|$ *consists of the components*

$$g_{pl} = w_l f_p + w_p f_l \quad \text{for all } p \in A, l \in B,$$
$$g_{lp} = \gamma_l f_p + \gamma_p f_l \quad \text{for all } p \in A, l \in B,$$
$$g_s = f_s \quad \text{for all } s \in I\backslash\{A \cup B\}.$$

In the second extreme case, the inequality $\frac{w_i}{\gamma_i} > \frac{w_j}{\gamma_j}$ takes place for a unique pair of indexes $i \in A, j \in B$. Then $P = \{p\}$, $L = \{l\}$, $P_l = P = \{p\}$, $\bar{P}_l = A\backslash\{p\}$, $L_p = L = \{l\}$, and $\bar{L}_p = B\backslash\{l\}$. Accordingly, we obtain the following result.

Theorem 5.3 *Assume that there is a given closed collection of two information quanta (A, B) where* $P = \{p\}$ *and* $L = \{l\}$. *Then for any set of selectable vectors* C(Y) *we have inclusions* (5.6), *where* $\hat{P}(Y) = f(P_g(X))$ *and the vector criterion g of dimension* $m - (|A| + |B|) + 2 + |\bar{P}_l| + |\bar{L}_p|$ *consists of the components*

$$g_{pl} = w_l f_p + w_p f_l,$$
$$g_{lp} = \gamma_l f_p + \gamma_p f_l,$$
$$g_{pli} = (\gamma_i w_l - \gamma_l w_i) f_p + (\gamma_l w_p - \gamma_p w_l) f_i + (\gamma_i w_p - \gamma_p w_i) f_l \quad \text{for all } i \in \bar{P}_l,$$
$$g_{plj} = (\gamma_l w_j - \gamma_j w_l) f_p + (\gamma_l w_p - \gamma_p w_l) f_j + (\gamma_p w_j - \gamma_j w_p) f_l \quad \text{for all } j \in \bar{L}_p,$$
$$g_s = f_s \quad \text{for all } s \in I\backslash\{A \cup B\}.$$

Example 5.1 Let $m = 10$ and $|A| = 5$ and $|B| = 5$. If there exists a unique pair of indexes $p \in A$, $l \in B$, under which inequality (5.1) holds, then the new vector criterion g has 10 components. Now, assume that inequality (5.1) takes place only for indexes $p_1, p_2 \in A$, $l \in B$. Then $|P_l| = 2$, $|L_{p_1}| = |L_{p_2}| = 1$, $|L_{p_2}| = 1$, $|\bar{P}_l| = |A| - |P_l| = 3$, $|\bar{L}_{p_1}| = |\bar{L}_{p_2}| = |B| - |L_{p_1}| = 4$ and, by formula (5.7), dimension of vector criterion g is equal to 18. If inequality (5.1) is true for all $i \in A$ and $j \in B$, then the new vector criterion g consists of 50 components.

It can be proved the following result.

Theorem 5.4 *Inclusions* (5.6) *are invariant with respect to a linear positive transformation of the components of the vector criterion* g *that are defined by formulas* (5.8).

5.2 Cyclic Collections of Information Quanta

5.2.1 Definition and Consistency of Cyclic Collection of Information Quanta

Introduce pairwise disjoint subsets $A_i \subset I, i = 1, 2, \ldots, k, \sum_{i=1}^{k} |A_i| \leq m$, and consider the following closed "chain" of information quanta:

(1) The group of criteria A_1 is more important than the group of criteria A_2 with parameters $w_{i_1}^{(1)}$ for all $i_1 \in A_1$ and $w_{i_2}^{(1)}$ for all $i_2 \in A_2$;
(2) The group of criteria A_2 is more important than the group of criteria A_3 with parameters $w_{i_2}^{(2)}$ for all $i_2 \in A_2$ and $w_{i_3}^{(2)}$ for all $i_3 \in A_3$; and so on.
(3) The group of criteria A_k is more important than the group of criteria A_1 with parameters $w_{i_k}^{(k)}$ for all $i_k \in A_k$ and $w_{i_1}^{(k)}$ for all $i_1 \in A_1$.

The corresponding formal definition can be found below.

Definition 5.2 We say that there is a given *cyclic collection of information quanta* with the groups of criteria $A_1, A_2, \ldots, A_k$ and the collection of positive parameters $\left\{ w_{i_1}^{(1)}, w_{i_2}^{(1)}, w_{i_2}^{(2)}, w_{i_3}^{(2)}, \ldots, w_{i_{k-1}}^{(k-1)}, w_{i_k}^{(k-1)}, w_{i_k}^{(k)}, w_{i_1}^{(k)} > 0 \text{ for all } i_1 \in A_1, i_2 \in A_2, \ldots, i_k \in A_k \right\}$ if the vectors $y^1, y^2, \ldots, y^k$ of the form

$$
\begin{aligned}
&y^1_{i_1} = w^{(1)}_{i_1}, \quad y^1_{i_2} = -w^{(1)}_{i_2}, \quad y^1_s = 0 \quad \text{for all} \quad i_1 \in A_1,\ i_2 \in A_2,\ s \in I\backslash(A_1 \cup A_2),\\
&y^2_{i_2} = w^2_{i_2}, \quad y^2_{i_3} = -w^{(2)}_{i_3}, \quad y^2_s = 0 \quad \text{for all} \quad i_2 \in A_2,\ i_3 \in A_3,\ s \in I\backslash(A_2 \cup A_3),\\
&\cdots\cdots\cdots\cdots\cdots\cdots\cdots\cdots\cdots\cdots\cdots\cdots\cdots\cdots\cdots\cdots\cdots\cdots\cdots\cdots\\
&y^k_{i_k} = w^{(k)}_{i_k}, \quad y^k_{i_1} = -w^{(k)}_{i_1}, \quad y^k_s = 0 \quad \text{for all} \quad i_k \in A_k,\ i_1 \in A_1,\ s \in I\backslash(A_k \cup A_1),
\end{aligned}
\tag{5.25}
$$

satisfy the relationships $y^1 \succ 0_m, y^2 \succ 0_m, \ldots, y^k \succ 0_m$.

For all $i_1 \in A_1$, $i_2 \in A_2$, ..., $i_k \in A_k$, introduce a square matrix $W(i_1, i_2, \ldots, i_k)$ of order k that depends on the criteria indexes $i_1, i_2, \ldots, i_k$ and has the form

$$
W(i_1, i_2, \ldots, i_k) = \begin{pmatrix}
w^{(1)}_{i_1} & 0 & \ldots & 0 & -w^{(k)}_{i_1}\\
-w^{(1)}_{i_2} & w^{(2)}_{i_2} & \ldots & 0 & 0\\
\ldots & \ldots & \ldots & \ldots & \ldots\\
0 & 0 & \ldots & w^{(k-1)}_{i_{k-1}} & 0\\
0 & 0 & \ldots & -w^{(k-1)}_{i_k} & w^{(k)}_{i_k}
\end{pmatrix}
$$

Cyclic collections of information quanta are mutually dependent and have some specifics. In this connection, the parameters of such collections may take values making them inconsistent. The next result gives a consistency criterion for cyclic information.

Theorem 5.6 *A cyclic collection of information quanta with the groups of criteria* $A_1, A_2, \ldots, A_k$ *is consistent if and only if there exist indexes* $i_p \in A_p, p = 1, 2, \ldots, k$, *such that the determinant of the matrix* $W(i_1, i_2, \ldots, i_k)$ *is positive, i.e.,* $|W(i_1, i_2, \ldots, i_k)| > 0$.

□ Clearly, the inequalities $|W(i_1, i_2, \ldots, i_k)| > 0$ and $\frac{w^{(1)}_{i_2} w^{(2)}_{i_3} \cdot \ldots \cdot w^{(k-1)}_{i_k} w^{(k)}_{i_1}}{w^{(1)}_{i_1} w^{(2)}_{i_2} \cdot \ldots \cdot w^{(k-1)}_{i_{k-1}} w^{(k)}_{i_k}} < 1$ are equivalent.

By the algebraic criterion of consistency from Chap. 4, a cyclic collection of information is consistent if and only if the system of linear equations

$$
\sum_{i=1}^{m} \lambda_i e^i + \sum_{j=1}^{k} \mu_j y^{(j)} = 0_m
$$

has no N-solution $(\lambda_1, \ldots, \lambda_m, \mu_1, \ldots, \mu_k)$, where the vectors $y^{(1)}$, $y^{(2)}$, ..., $y^{(k)}$ are of form (5.25). In other words, there exist indexes $i'_p \in A_p$, $p = 1, 2, \ldots, k$, such that the system of linear equations $\omega(i'_1, i'_2, \ldots, i'_k) W_E(i'_1, i'_2, \ldots, i'_k) = 0_k$ has no solution $\omega(i'_1, i'_2, \ldots, i'_k) \geq 0_{2k}$, where

$$
\begin{aligned}
\omega(i_1, i_2, \ldots, i_k) &= (\lambda_{i_1}, \lambda_{i_2}, \ldots, \lambda_{i_k}, \mu_1, \mu_2, \ldots, \mu_k),\\
W_E(i_1, i_2, \ldots, i_k) &= \begin{pmatrix} E_{k\times k} \\ W^T(i_1, i_2, \ldots, i_k) \end{pmatrix},
\end{aligned}
$$

and $E_{k\times k}$ is the identity matrix of order k. According to Motzkin's theorem of the alternative (e.g., see [25]), the last statement is equivalent to the existence of a solution $x \in R^k$ to the system of inequalities $W_E(i'_1, i'_2, \ldots, i'_k)x > 0_{2k}$.

Therefore, a cyclic collection of information quanta is consistent if and only if there exist criteria indexes $i'_p \in A_p$, $p = 1, 2, \ldots, k$, such that the system of linear inequalities $W_E(i'_1, i'_2, \ldots, i'_k)x > 0_{2k}$ has some solution.

Necessity. Assume that there exist indexes $i'_p \in A_p$, $p = 1, 2, \ldots, k$, and a vector $x \in R^k$ such that $W_E(i'_1, i'_2, \ldots, i'_k)x > 0_{2k}$. Then

$$\begin{aligned} x &> 0_k, \\ w_{i'_1}^{(1)} x_1 - w_{i'_2}^{(1)} x_2 &> 0, \\ &\cdots \\ w_{i'_{k-1}}^{(k-1)} x_{k-1} - w_{i'_k}^{(k-1)} x_k &> 0, \\ w_{i'_k}^{(k)} x_k - w_{i'_1}^{(k)} x_1 &> 0. \end{aligned} \tag{5.26}$$

Trivial transformations applied to (5.26) yield the inequality

$$\frac{w_{i'_2}^{(1)} w_{i'_3}^{(2)} \cdot \ldots \cdot w_{i'_k}^{(k-1)} w_{i'_1}^{(k)}}{w_{i'_1}^{(1)} w_{i'_2}^{(2)} \cdot \ldots \cdot w_{i'_{k-1}}^{(k-1)} w_{i'_k}^{(k)}} x_1 < x_1.$$

And the desired result follows by dividing both sides by $x_1 > 0$.

Sufficiency. Assume that there exist indexes $i'_p \in A_p$, $p = 1, 2, \ldots, k$, such that the inequality $\frac{w_{i'_2}^{(1)} w_{i'_3}^{(2)} \cdot \ldots \cdot w_{i'_k}^{(k-1)} w_{i'_1}^{(k)}}{w_{i'_1}^{(1)} w_{i'_2}^{(2)} \cdot \ldots \cdot w_{i'_{k-1}}^{(k-1)} w_{i'_k}^{(k)}} < 1$ holds. Multiply it by an arbitrary positive number x_1 to obtain

$$\frac{w_{i'_1}^{(1)} w_{i'_2}^{(2)} \cdot \ldots \cdot w_{i'_{k-1}}^{(k-1)}}{w_{i'_2}^{(1)} w_{i'_3}^{(2)} \cdot \ldots \cdot w_{i'_k}^{(k-1)}} x_1 > \frac{w_{i'_1}^{(k)}}{w_{i'_k}^{(k)}} x_1.$$

Obviously, we can find a number x_k such that

$$\frac{w_{i'_1}^{(1)} w_{i'_2}^{(2)} \cdot \ldots \cdot w_{i'_{k-1}}^{(k-1)}}{w_{i'_2}^{(1)} w_{i'_3}^{(2)} \cdot \ldots \cdot w_{i'_k}^{(k-1)}} x_1 > x_k > \frac{w_{i'_1}^{(k)}}{w_{i'_k}^{(k)}} x_1,$$

whence it appears that $\frac{w_{i'_1}^{(1)} w_{i'_2}^{(2)} \cdot \ldots \cdot w_{i'_{k-2}}^{(k-2)}}{w_{i'_2}^{(1)} w_{i'_3}^{(2)} \cdot \ldots \cdot w_{i'_{k-1}}^{(k-2)}} x_1 > \frac{w_{i'_k}^{(k-1)}}{w_{i'_{k-1}}^{(k-1)}} x_k$. Next, there esists a number x_{k-1} such that

$$\frac{w_{i_1'}^{(1)} w_{i_2'}^{(2)} \cdot \ldots \cdot w_{i_{k-2}'}^{(k-2)}}{w_{i_2'}^{(1)} w_{i_3'}^{(2)} \cdot \ldots \cdot w_{i_{k-1}'}^{(k-2)}} x_1 > x_{k-1} > \frac{w_{i_k'}^{(k-1)}}{w_{i_{k-1}'}^{(k-1)}} x_k.$$

Subsequently,

$$\frac{w_{i_1'}^{(1)} \cdot \ldots \cdot w_{i_{k-3}'}^{(k-3)}}{w_{i_2'}^{(1)} \cdot \ldots \cdot w_{i_{k-2}'}^{(k-3)}} x_1 > \frac{w_{i_{k-1}'}^{(k-2)}}{w_{i_{k-2}'}^{(k-2)}} x_{k-1}.$$

Using the same line of reasoning, we demonstrate that there exists a number x_2 such that

$$\frac{w_{i_1'}^{(1)}}{w_{i_2'}^{(1)}} x_1 > x_2 > \frac{w_{i_3'}^{(2)}}{w_{i_2'}^{(2)}} x_3.$$

And trivial transformations of these inequalities give (5.26). Moreover, since $x_1 > 0$, we have $x > 0_k$. ■

5.2.2 *Pareto Set Reduction Based on Cyclic Collections of Information Quanta*

This subsection deals with Pareto set reduction using cyclic collections of information quanta. First, consider the elementary situation, i.e., all groups in the definition of cyclic information represent singletons: $A_s = \{i_s\}$, $s = 1, 2, \ldots, k$. In this case, we have *cyclic information with criteria (criteria indexes)* i_1, i_2, ..., i_k, and denote by W the corresponding matrix $W(i_1, i_2, \ldots, i_k)$. And the necessary and sufficient condition of consistency (Theorem 5.6) is reduced to the inequality $|W| > 0$.

Introduce the family of matrices $W_{s,p}, s = 1, 2, \ldots, k, p = 1, 2, \ldots, k,$ where column s of the matrix W (further designated by W^s) is the unit vector e^p of space R^k, i.e.,

$$W_{s,p} = \left(W^1, \ldots, W^{s-1}, e^p, W^{s+1}, \ldots, W^k\right).$$

Lemma 5.3 *The determinant of the matrix* $W_{s,p}$ *is positive for all* $s = 1, 2, \ldots, k, p = 1, 2, \ldots, k$.

□ Let M_{ij} be the $(k - 1)$-th minor of the matrix W obtained after elimination of row i and column j from this matrix. In addition, consider the two triangular matrices

$$L(l,q)=\begin{pmatrix} w_{i_l}^{(l)} & 0 & \dots & 0 & 0\\ -w_{i_{l+1}}^{(l)} & w_{i_{l+1}}^{(l+1)} & \dots & 0 & 0\\ \dots & \dots & \dots & \dots & \dots\\ 0 & 0 & \dots & w_{i_{q-1}}^{(q-1)} & 0\\ 0 & 0 & \dots & -w_{i_q}^{(q-1)} & w_{i_q}^{(q)} \end{pmatrix},$$

where $l, q \in \{1, 2, \dots, k\}$, $l \leqq q$, and

$$U(r,t)=\begin{pmatrix} -w_{i_{r+1}}^{(r)} & w_{i_{r+1}}^{(r+1)} & \dots & 0 & 0\\ 0 & -w_{i_{r+2}}^{(r+1)} & \dots & 0 & 0\\ \dots & \dots & \dots & \dots & \dots\\ 0 & 0 & \dots & -w_{i_{t-1}}^{(t-2)} & w_{i_{t-1}}^{(t-1)}\\ 0 & 0 & \dots & 0 & -w_{i_t}^{(t-1)} \end{pmatrix},$$

where $r, t \in \{1, 2, \dots, k\}$, $r<t$. Construct a matrix $Z_{(n\times m)}$ of dimensions $n \times m$ where all elements except the last element $-w_{i_1}^{(k)}$ in row 1 are zero. The determinants of these triangular matrices make the product of their diagonal elements. Therefore, the determinant of the matrix $L(l, q)$ is positive for all possible values of l and q. Denote it by $\alpha(l, q)$. The sign of the determinant of the matrix $U(r, t)$ depends on the difference $(t-r)$, since $|U(r,t)| = (-1)^{t-r}\beta(r,t)$, where $\beta(r, t)$ is some positive number.

We will calculate the determinant $|W_{s,p}|$ using the column expansion with respect to column s:

$$|W_{s,p}| = (-1)^{s+p} M_{sp} \tag{5.27}$$

The $(k-1)$-th minor M_{sp} varies depending on the relationship between the numbers s and p. Consider three cases as follows: (1) $s = p$; (2) $s > p$; (3) $s<p$.

(1) $s = p$. The power of (-1) in (5.39) is $2s$, and the sign of $|W_{s,s}|$ is defined by the sign of M_{ss}. If $s = 1$ or $s = k$, then

$$M_{11} = |L(2,k)| = \alpha(2,k), \quad M_{kk} = |L(1,k-1)| = \alpha(1,k-1).$$

Let $s \neq 1$ and $s \neq k$. Then

$$M_{ss} = \begin{vmatrix} L(1,s-1) & Z_{(s-1)\times(k-s)}\\ 0_{(k-s)\times(s-1)} & L(s+1,k) \end{vmatrix} = \alpha(1,s-1)\alpha(s+1,k).$$

(2) $s > p$. If $p = 1, s \neq k$ or $s = k, p \neq 1$, we accordingly obtain

$$M_{s1} = \begin{vmatrix} U(1,s) & 0_{(s-1)\times(k-s)} \\ 0_{(k-s)\times(s-1)} & L(s+1,k) \end{vmatrix} = (-1)^{s-1}\,\alpha(s+1,k)\,\beta(1,s),$$

$$M_{kp} = \begin{vmatrix} L(1,p-1) & 0_{(p-1)\times(k-p)} \\ 0_{(k-p)\times(p-1)} & U(p,k) \end{vmatrix} = (-1)^{k-p}\,\alpha(1,p-1)\,\beta(p,k).$$

For $p = 1$ and $s = k$, the result is $M_{k1} = |U(1,k)| = (-1)^{k-1}\beta(1,k)$. Now, let $s \neq k$ and $p \neq 1$. Subsequently,

$$M_{sp} = \begin{vmatrix} L(1,p-1) & 0_{(p-1)\times(s-p)} & Z_{(p-1)\times(k-s)} \\ 0_{(s-p)\times(p-1)} & U(p,s) & 0_{(s-p)\times(k-s)} \\ 0_{(k-s)\times(p-1)} & 0_{(k-s)\times(s-p)} & L(s+1,k) \end{vmatrix}$$
$$= (-1)^{p-s}\,\alpha(1,p-1)\,\beta(p,s)\,\alpha(s+1,k).$$

Based on the above expressions for the minor M_{sp} and (5.27), we conclude that the determinant $|W_{s,p}|$ is positive in the case $s > p$.

(3) $s<p$. If $s = 1$ and $p \neq k$, then

$$M_{1p} = \begin{vmatrix} 0 \;\; 0 & \dots & 0 & -w_{i_1}^{(k)} \\ & & & 0 \\ L(2,p-1) & 0_{(p-2)\times(k-p)} & & \vdots \\ 0_{(k-p)\times(p-2)} & U(p,k) & & 0 \\ & & & w_{i_k}^{(k)} \end{vmatrix} = (-1)^{2k-p+1} w_{i_1}^{(k)}\alpha(2,p-1)\,\beta(p,k).$$

For $p = k$ and $s \neq 1$, it follows that

$$M_{sk} = \begin{vmatrix} w_{i_1}^{(1)} \;\; 0 & \dots & 0 & -w_{i_1}^{(k)} \\ & & & 0 \\ U(1,s) & 0_{(s-1)\times(k-s-1)} & & \vdots \\ 0_{(k-s-1)\times(s-1)} & L(s+1,k-1) & & 0 \\ & & & 0 \end{vmatrix}$$
$$= (-1)^{k+s} w_{i_1}^{(k)}\alpha(s+1,k-1)\beta(1,s).$$

With $s = 1$ and $p = k$, we have

$$M_{1k} = \begin{vmatrix} 0 & 0 & \dots & 0 & -w_{i_1}^{(k)} \\ & & & & 0 \\ & L(2,k-1) & & & \vdots \\ & & & & 0 \\ & & & & 0 \end{vmatrix} = (-1)^{k+1} w_{i_1}^{(k)} \alpha(2,k-1).$$

Now, let $s \neq 1$ and $p \neq k$. The minor obtained from the minor M_{sp} by eliminating the row and column that contain $w_{i_1}^{(1)}$ is zero. Using this and performing the raw expansion of minor M_{sp} with respect to row 1, we obtain

$$M_{sp} = \begin{vmatrix} w_{i_1}^{(1)} \quad 0 & \dots \quad \dots & \dots \quad 0 & -w_{i_1}^{(k)} \\ & & & 0 \\ U(1,s) & 0_{(s-1)\times(p-s-1)} & 0_{(s-1)\times(k-p)} & \vdots \\ 0_{(p-s-1)\times(s-1)} & L(s+1,p-1) & 0_{(p-s-1)\times(k-p)} & 0 \\ 0_{(k-p)\times(s-1)} & 0_{(k-p)\times(p-s-1)} & U(p,k) & 0 \end{vmatrix}$$

$$= (-1)^{2k+s-p} w_{i_1}^{(k)} \alpha(s+1,p-1)\, \beta(1,s)\, \beta(p,k).$$

Based on these results and formula (5.27), a conclusion is that the determinant $|W_{s,p}|$ takes a positive value in the case.$s<p$.

Therefore, we have established that the determinant $|W_{s,p}|$ is positive in all the three cases $(s = p, s > p, s<p)$. ■

Theorem 5.7 *Assume that there is a given consistent cyclic collection of information with criteria* $i_1, i_2, \dots, i_k$ *and a corresponding collection of positive parameters. Then for any set of selectable vectors* $C(Y)$ *we have inclusions* (5.6), *where* $\hat{P}(Y) = f(P_g(X))$ *and the "new" m-dimensional vector criterion g has the components*

$$g_{i_s} = \sum_{p=1}^{k} |W_{s,p}| f_{i_p} \quad s = 1, 2, \dots, k; \quad g_i = f_i, \quad \textit{for all } i \in I \backslash I_k. \tag{5.28}$$

□ I. Let K be the acute convex cone (without the origin) of the cone relation $\succ$. The specification of a cyclic collection of information means that the vectors $y^{(1)}, y^{(2)}, \dots, y^{(k)}$ with the components

$$\begin{aligned} y_{i_1}^1 &= w_{i_1}^{(1)}, \quad y_{i_2}^1 = -w_{i_2}^{(1)}, \quad y_s^1 = 0 \quad \text{for all} \quad s \in I\backslash(i_1 \cup i_2), \\ y_{i_2}^2 &= w_{i_2}^{(2)}, \quad y_{i_3}^2 = -w_{i_3}^{(2)}, \quad y_s^2 = 0 \quad \text{for all} \quad s \in I\backslash(i_2 \cup i_3), \\ &\dots\dots\dots\dots\dots\dots\dots\dots\dots\dots\dots\dots \\ y_{i_k}^k &= w_{i_k}^{(k)}, \quad y_{i_1}^k = -w_{i_1}^{(k)}, \quad y_s^k = 0 \quad \text{for all} \quad s \in I\backslash(i_k \cup i_1), \end{aligned}$$

satisfy the relationships $y^1 \succ 0_m$, $y^2 \succ 0_m$, ..., $y^k \succ 0_m$. This is equivalent to the inclusions $y^1, y^2, \ldots, y^k \in K$.

Let M be the acute convex cone (without the origin) generated by the vector collection

$$e^s \quad \text{for all} \quad s \in I \backslash I_k, \quad y^1, y^2, \ldots, y^k.$$

Show that all vectors from this collection are the generators of the cone M. Conjecture the opposite. Let a certain unit vector e^s, $s \notin I_k$, be representable as the following N-combination

$$e^s = \sum_{l \notin I_k \cup \{s\}} \lambda_l e^l + \sum_{i=1}^{k} \mu_i y^i.$$

The above linear combination fails for any coefficients, since component s of the last equality is $1 = 0$.

Assume that there exists $p \in \{1, 2, \ldots, k\}$ such that

$$y^p = \sum_{l \notin I_k} \lambda_l e^l + \sum_{\substack{i=1 \\ i \neq p}}^{k} \mu_i y^i. \tag{5.29}$$

If $p = 1$, for component i_l of (5.29) we have $w_{i_1}^{(1)} = -\mu_k w_{i_1}^{(k)}$. And if $p \in \{2, \ldots, k\}$, then component i_p of the vector equality (5.29) is $w_{i_p}^{(p)} = -\mu_{p-1} w_{i_p}^{(p-1)}$. The both last equalities are impossible due to nonnegativity of μ_k and μ_{p-1}. Thus, the vector equality (5.29) takes no place.

Show that the unit vectors e^{i_s}, $s = 1, 2, \ldots, k$, belong to the cone M, i.e., each of them is the following N-combination:

$$e^{i_s} = \sum_{l \notin I_k} \lambda_l^{(s)} e^l + \sum_{i=1}^{k} \mu_i^{(s)} y^i, \quad s = 1, 2, \ldots, k.$$

Choose $\lambda_l^{(s)} = 0$ for $s = 1, 2, \ldots, k$ and any $l \in I \backslash I_k$. The unknowns $\mu_p^{(s)}$, $s = 1, 2, \ldots, k, p = 1, 2, \ldots, k$, can be found as the solutions of the system of linear equations

$$W\mu^{(s)} = \hat{e}^s, \quad s = 1, 2, \ldots, k, \tag{5.30}$$

where $\mu^{(s)} = \left(\mu_1^{(s)}, \mu_2^{(s)}, \ldots, \mu_k^{(s)}\right)^T$ and $\hat{e}^s \in R^k$. The inequality $|W| > 0$ holds due to the consistency of the cyclic collection of information. Hence, system (5.30) possesses a unique solution for all $s = 1, 2, \ldots, k$. Using Cramer's rule, we obtain

$$\mu_p^{(s)} = \frac{|W_{p,s}|}{|W|} > 0,$$

since $|W| > 0$ and $|W_{s,p}| > 0$ for all $s = 1,2,\ldots,k, p = 1,2,\ldots,k$ (see Lemma 5.3).

In fact, we have established that the generators of the cone M are the vectors e^s for all $s \in I \backslash I_k$, $y^1, y^2, \ldots, y^k$, and also that the inclusions $e^{i_s} \in M, s = 1,2,\ldots,k$, hold.

II. Now, it is necessary to prove that the cone M coincides with the nonzero solutions to the system of linear inequalities

$$\begin{array}{c} \langle e^l, y \rangle \geqq 0 \quad \text{for all} \quad l \in I \backslash I_k, \\ \langle \bar{y}^s, y \rangle \geqq 0 \quad s = 1,2,\ldots,k, \\ \bar{y}^s_{i_p} = |W_{s,p}|, \; p = 1,2,\ldots,k; \quad \bar{y}^s_q = 0 \quad \text{for all } q \in I \backslash I_k, \quad s = 1,2,\ldots,k. \end{array} \tag{5.31}$$

Find the fundamental system of solutions to the system of inequalities (5.31). The idea is to show that the latter coincides with the vector collection e^s for all $s \in I \backslash I_k$ and $y^1, y^2, \ldots, y^k$.

Consider the system of linear equations that correspond to inequalities (5.31), i.e.,

$$\begin{array}{l} \langle e^l, y \rangle = 0 \quad \text{for all} \quad l \in I \backslash I_k, \\ \langle \bar{y}^s, y \rangle = 0, \quad s = 1,2,\ldots,k. \end{array} \tag{5.32}$$

The matrix of any subsystem obtained from (5.32) by eliminating a certain equation has rank $(m-1)$. To get the fundamental system of solutions, it suffices to find the nonzero solutions to all possible subsystems of $(m-1)$ equations from system (5.32) that satisfy the system of inequalities (5.31).

If the equation $\langle e^l, y \rangle = 0$ is removed from (5.32), then the resulting "truncated" subsystem has the solution e^l that satisfies (5.31). If we eliminate the equation $\langle \bar{y}^p, y \rangle = 0$, the solution is the vector $y^p, p = 1,2,\ldots,k$. Let us demonstrate it. Obviously, $\langle \bar{y}^p, e^l \rangle = 0$ for all $l \in I \backslash I_k$. Consider the scalar product $\langle \bar{y}^s, y^p \rangle$, $s = 1,2,\ldots,k$, $s \neq p$. Our analysis will be confined to the case $p \neq k$ (if $p = k$, the line of reasoning is the same).

Here three cases are possible, namely, (1) $s = 1$; (2) $s = k$; (3) $s \neq 1$ and $s \neq k$.

(1) Let $s = 1$. In this case, we have

$$\begin{aligned} \langle \bar{y}^1, y^p \rangle &= |W_{1,p}| \cdot w^{(p)}_{i_p} - |W_{1,p+1}| \cdot w^{(p)}_{i_{p+1}} = |\hat{e}^p, W^2, \ldots, W^k| \cdot w^{(p)}_{i_p} - |\hat{e}^{p+1}, W^2, \ldots, W^k| \cdot w^{(p)}_{i_{p+1}} \\ &= \left| w^{(p)}_{i_p} \hat{e}^p - w^{(p)}_{i_{p+1}} \hat{e}^{p+1}, W^2, \ldots, W^k \right| = |W^p, W^2, \ldots, W^k| = 0, \end{aligned}$$

since the columns $W^2, \ldots, W^k$ surely contain the column W^p.

(2) Let $s = k$. Then

$$\langle \bar{y}^k, y^p \rangle = |W_{k,p}| \cdot w_{i_p}^{(p)} - |W_{k,p+1}| \cdot w_{i_{p+1}}^{(p)} = |W^1, \ldots, W^{k-1}, \hat{e}^{(p)}| \cdot w_{i_p}^{(p)} - |W^1, \ldots, W^{k-1}, \hat{e}^{p+1}| \cdot w_{i_{p+1}}^{(p)} = |W^1, \ldots, W^{k-1}, W^p| = 0,$$

since the columns $W^1,\ldots, W^{k-1}$ surely contain the column W^p.

3) Let $s \neq 1$ and $s \neq k$. It follows that

$$\begin{aligned}\langle \bar{y}^s, y^p \rangle &= |W_{s,p}| \cdot w_{i_p}^{(p)} - |W_{s,p+1}| \cdot w_{i_{p+1}}^{(p)} \text{Ë} \\ &- |W^1, \ldots, W^{s-1}, \hat{e}^{p+1}, W^{s+1}, \ldots, W^k| \cdot w_{i_{p+1}}^{(p)} \\ &= |W^1, \ldots, W^{s-1}, W^p, W^{s+1}, \ldots, W^k| = 0,\end{aligned}$$

since the columns $W^1, \ldots, W^{s-1}, W^{s+1}, \ldots, W^k$ surely contain the column W^p.

For checking that $y^{(p)}$ satisfies (5.31), it remains to show that $\langle \bar{y}^p, y^p \rangle \geqq 0$. If $p = 1$, we have

$$\begin{aligned}\langle \bar{y}^1, y^1 \rangle &= |W_{1,1}| \cdot w_{i_1}^{(1)} - |W_{1,2}| \cdot w_{i_2}^{(1)} \\ &= |\hat{e}^1, W^2, \ldots, W^k| \cdot w_{i_1}^{(1)} - |\hat{e}^2, W^2, \ldots, W^k| \cdot w_{i_2}^{(p)} = |W| > 0.\end{aligned}$$

For $p = k$,

$$\begin{aligned}\langle \bar{y}^k, y^k \rangle &= |W_{k,k}| \cdot w_{i_k}^{(k)} - |W_{k,1}| \cdot w_{i_1}^{(k)} \\ &= |W^1, \ldots, W^{k-1}, \hat{e}^k| \cdot w_{i_k}^{(k)} - |W^1, \ldots, W^{k-1}, \hat{e}^1| \cdot w_{i_1}^{(k)} = |W| > 0.\end{aligned}$$

With $p \neq 1$ and $p \neq k$, the result is

$$\begin{aligned}\langle \bar{y}^p, y^p \rangle &= |W_{p,p}| \cdot w_{i_p}^{(p)} - |W_{p,p+1}| \cdot w_{i_{p+1}}^{(p)} \\ &= \left|W^1, \ldots, W^{p-1}, w_{i_p}^{(p)} \cdot \hat{e}^{(p)} - w_{i_{p+1}}^{(p)} \cdot \hat{e}^{(p+1)}, W^{p+1}, \ldots, W^k\right| \\ &= |W^1, \ldots, W^{p-1}, W^p, W^{p+1}, \ldots, W^k| = |W| > 0.\end{aligned}$$

Therefore, the fundamental system of solutions to the system of inequalities (5.31) is the vector collection e^s for all $s \in I \backslash I_k$ and $y^1, y^2, \ldots, y^k$. Hence the cone M coincides with the set of the nonzero solutions to system (5.31).

III. The inclusions $R_+^m \subset M \subset K$ imply

$$\text{Ndom}\, Y \subset \hat{P}(Y) \subset P(Y),$$

where

$$\hat{P}(Y) = \{y^* \in Y| \quad \text{there exists no } y \in Y \quad \text{such that } y - y^* \in M\}.$$

The remainder of the proof is similar to Theorem 5.1 and is therefore omitted. ■

Corollary 5.1 *Under the hypotheses of Theorem 5.7, choose* $k = 2$, $i_1 = i$, *and* $i_2 = j$. *Then formulas* (5.28) *for the "new" vector criterion g are reduced to*

$$g_i = |W_{1,1}|f_i + |W_{1,2}|f_j = w_j^{(2)}f_i + w_i^{(2)}f_j,$$
$$g_j = |W_{2,1}|f_i + |W_{2,2}|f_j = w_j^{(1)}f_i + w_i^{(1)}f_j,$$
$$g_s = f_s \quad \textit{for all } s \in I\backslash\{i,\ j\},$$

$$|W_{1,1}| = \begin{vmatrix} 1 & -w_i^{(2)} \\ 0 & w_j^{(2)} \end{vmatrix} = w_j^{(2)}, \quad |W_{1,2}| = \begin{vmatrix} 0 & -w_i^{(2)} \\ 1 & w_j^{(2)} \end{vmatrix} = w_i^{(2)},$$

$$|W_{2,1}| = \begin{vmatrix} w_i^{(1)} & 1 \\ -w_j^{(1)} & 0 \end{vmatrix} = w_j^{(1)}, \quad |W_{2,2}| = \begin{vmatrix} w_i^{(1)} & 0 \\ -w_j^{(1)} & 1 \end{vmatrix} = w_i^{(1)}.$$

Corollary 5.2 *Under the hypotheses of Theorem 5.7, choose* $k = 3$, $i_1 = i$, $i_2 = j$, *and* $i_3 = l$. *Then formulas* (5.28) *for the "new" vector criterion g are reduced to*

$$g_i = w_j^{(2)}w_l^{(3)}f_i + w_l^{(2)}w_i^{(3)}f_j + w_j^{(2)}w_i^{(3)}f_l,$$
$$g_j = w_j^{(1)}w_l^{(3)}f_i + w_i^{(1)}w_l^{(3)}f_j + w_j^{(1)}w_i^{(3)}f_l,$$
$$g_l = w_j^{(1)}w_l^{(2)}f_i + w_i^{(1)}w_l^{(2)}f_j + w_i^{(1)}w_j^{(2)}f_l,$$
$$g_s = f_s \quad \textit{for all} \quad s \in I\backslash\{i, j, l\}.$$

5.3 Geometrical Algorithm of New Vector Criterion Design

5.3.1 Preliminary Analysis

Clearly, the above cases of using different collections of information quanta do not exhaust all possible situations. Of course, this applies to the mutually dependent collections.

Let us discover a common scheme for obtaining the recalculation formulas of the new vector criterion, as appears from the proofs of all theorems on taking into account different information about the DM's preference relation. This scheme can be described in brief in the following way. Right from the start (i.e., when there exists no information in the form of quanta), only the inclusion $R_+^m \subset K$ holds owing to the Pareto axiom; here K denotes the acute convex cone of the cone relation $\succ$. Generally speaking, the availability of a certain collection of k information quanta in geometrical terms means the specification of k vectors y^i with at least one positive and at least one negative componnets such that $y^i \succ 0_m$ or, equivalently, $y^i \in K$, $i = 1, 2, \ldots, k$. Next, we introduce the acute convex cone M (without the origin) generated by the vectors $e^1, e^2, \ldots, e^m, y^1, y^2, \ldots, y^k$. This cone defines a wider cone relation from the same class as the unknown preference relation $\succ$, since $M \subset K$. The cone M is finitely generated and therefore polyhedral. The dimension of the new vector criterion coincides with the number of the $(m-1)$-dimensional facets of the cone M, while the inner normals of these facets yield the recalculation formulas of the new vector criterion.

For example, in the elementary case (criterion i is more important than criterion j with parameters w_i, w_j), the cone M has the inner normals $e^1, \ldots, e^{j-1}, w_j e^i + w_i e^j, e^{j+1}, \ldots, e^m$, see proof of Theorem 2.5. And the resulting recalculation formula of the new vector criterion j is given by $g_j = w_j f_i + w_i f_j$.

According to the aforesaid, we are going to consider the general problem whose solution would yield the recalculation formulas of the new criteria in any situations with arbitrary collections of information quanta.

Recall the definition of a dual cone. Let $a^1, a^2, \ldots, a^{m+k}$ be a finite collection of vectors in the m-dimensional Euclidean space. Denote by

$$M = cone\{a^1, a^2, \ldots, a^{m+k}\}$$

the convex (polyhedral) cone generated by the above vectors. It consists of all nonnegative linear combinations of the vectors $a^1, a^2, \ldots, a^{m+k}$. By assumption, the cone is acute and m-dimensional.[1]

Let M^o be *the dual cone* for M, i.e.,

$$M^o = \{x \in R^m | \langle x, y \rangle \geq 0 \quad \text{for all}\, y \in M\}.$$

The dual cone for a polyhedral (finitely generated) cone also forms a polyhedral cone, thereby being generated by a certain finite collection of vectors. Another well-known result [57] states that the dual cone for an acute m-dimensional cone is acute and m-dimensional. And so, the consideration of an arbitrary finite collection of information quanta about the DM's preference relation can be reduced to the following problem.

Problem

Given a finite collection of vectors $a^1, a^2, \ldots, a^{m+k}$ that generate an acute polyhedral m-dimensional cone M, develop an algorithm yielding the minimal collection of vectors $b^1, b^2, \ldots, b^n$ that generate the dual cone M^o, i.e.,

$$M^o = cone\{b^1, b^2, \ldots, b^n\}.$$

[1]*The dimension of a cone* coincides with the dimension of the minimal subspace containing it.

In geometrical terms, this problem is to construct the collection of normal vectors for all hyperplanes representing the $(m - 1)$-dimensional facets of a cone M using the generators of M.

Note that the problem becomes trivial in the special case where M is the nonnegative orthant of space R^m_+. One of the solutions is the collection of the unit vectors of this space (exactly the ones generating the nonnegative orthant under consideration).

For any finite consistent collection of information quanta, the above algorithm can be used to obtain the recalculation formulas of the new vector criterion from the old one. And the resulting new criterion is then applied to construct an upper estimate for the set of selectable alternatives (vectors) in the associated multicriteria choice problem.

5.3.2 Geometrical Algorithm and Its Justification

This subsection provides the so-called *geometrical algorithm* for new vector criterion design using an arbitrary finite collection of consistent information quanta.

As the input data, the algorithm requires a finite collection of vectors $a^1, a^2, \ldots, a^{m+k}$, $k \geqq 1$, that generate an acute m-dimensional convex (polyhedral) cone M in space R^m. As the output data, the algorithm forms a new collection of vectors $b^1, b^2, \ldots, b^n$ that generate the dual cone in the same space.

Step 1 (looping for all possible vectors). Open a loop on variable i from 1 to C^{m-1}_{m+k} to generate all possible subcollections of $(m - 1)$ vectors from the collection $a^1, a^2, \ldots, a^{m+k}$.

Step 2 (checking of linear independence). If a current i-th subcollection $a^{i1}, \ldots, a^{i(m-1)}$ selected from $a^1, \ldots, a^{m+k}$ is linear dependent, set $i = i + 1$ and again execute Step 2. If further increment of i is impossible (i.e., $i = C^{m-1}_{m+k}$), move to Step 5. Otherwise (i.e., the above subcollection is linearly independent), go to Step 3.

Step 3 (constructing the orthogonal vector for the linearly independent subcollection). Take the column vectors of the subcollection $a^{i1}, \ldots, a^{i(m-1)}$ and enlarge them to a square matrix D of order n as follows. Attach to the right any vector from the set $I_i = \{a^1, \ldots, a^{m+k}\} \backslash \{a^{i1}, \ldots, a^{i(m-1)}\}$ that forms a linearly independent system together with $a^{i1}, \ldots, a^{i(m-1)}$ (such a vector surely exists, since the cone M is m-dimensional). Find the last column of the inverse matrix $(D^T)^{-1}$, where T denotes matrix transposition. Memorize this vector column and denote it by z^i. By the construction procedure, the vector z^i is orthogonal to all vectors from the subcollection $a^{i1}, \ldots, a^{i(m-1)}$.

Step 4 (checking of the belonging of z^i to the desired collection $b^1, b^2, \ldots, b^n$). Calculate the scalar products $\langle a^j, z^i \rangle$ for all vectors $a^j \in I_i$. If at least one scalar product is negative, then eliminate the vector z^i from memory. Otherwise (i.e., all scalar products are nonnegative), set $i = i + 1$ and get back to Step 2. If further increment of i is impossible, move to Step 5.

Step 5 (completing the calculations). Looping on variable i yields the column vectors memorized as z^i. These vectors form the desired minimal collection of the vectors $b^1, b^2, \ldots, b^n$ generating the dual cone M^o.

Let us justify the algorithm. We begin with an auxiliary result.

Lemma 5.4 [61] *A vector $z^i \in M^o$ is a generator*[2] *of the dual cone M^o if and only if the collection $a^1, a^2, \ldots, a^{m+k}$ contains $(m-1)$ linearly independent vectors with zero scalar products by the vector z^i and $(k+1)$ vectors with nonnegative scalar products by the vector z^i.*

□ Sufficiency. Denote by $a^{i1}, \ldots, a^{i(m-1)}$ the linearly independent vectors. By the hypothesis, we have

$$\begin{aligned} &\langle a^{ij}, z^i \rangle = 0, \quad j = 1, \ldots, m-1, \\ &\langle a^{ij}, z^i \rangle \geqq 0, \quad j = m, \ldots, m+k. \end{aligned} \tag{5.33}$$

Suppose that the vector z^i is not a generator of the dual cone. In this case, there exist two noncollinear vectors $t^1, t^2 \in M^o$ and two positive numbers α_1, α_2 such that $y^i = \alpha_1 t^1 + \alpha_2 t^2$. Since $t^1, t^2 \in M^o$,

$$\langle a^{ij}, t^1 \rangle \geqq 0, \quad \langle a^{ij}, t^2 \rangle \geqq 0, \quad j = 1, \ldots, m-1. \tag{5.34}$$

It follows from the equalities in (5.33) that

$$\langle a^{ij}, z^i \rangle = \alpha_1 \langle a^{ij}, t^1 \rangle + \alpha_2 \langle a^{ij}, t^2 \rangle = 0, \quad j = 1, \ldots, m-1.$$

Due to $\alpha_1, \alpha_2 > 0$ and inequalities (5.34), we obtain the relationships $\alpha_1 \langle a^{ij}, t^1 \rangle = \alpha_2 \langle a^{ij}, t^2 \rangle = 0$, $j = 1, \ldots, m-1$. This means that the linear hull $L(t^1, t^2)$ of the vectors t^1, t^2 is contained in the orthogonal complement of the linear hull of the vectors $a^{i1}, \ldots, a^{i(m-1)}$, i.e., $L(t^1, t^2) \subset L^{\perp}(a^{i1}, \ldots, a^{i(m-1)})$ and $\dim L(t^1, t^2) = 2$. Hence, $\dim L^{\perp}(a^{i1}, \ldots, a^{i(m-1)}) \geqq 2$, which is inconsistent with the linear independence of the subcollection of the m-dimensional vectors $a^{i1}, \ldots, a^{i(m-1)}$.

Necessity. Let a vector z^i be a generator of the dual cone M^o. Obviously, this vector satisfies the inequalities $\langle a^j, z^i \rangle \geqq 0, \quad j = 1, \ldots, m+k$. According to [62],

[2] *A generator* (*edge*) of a polyhedral cone is a vector of the cone that is not representable in the form of a positive linear combination of two other vectors belonging to this cone (see Subcection 2.1.2). All edges form the minimal system of vectors generating a given cone.

the generators of the acute m-dimensional cone M^o and the $(m-1)$-dimensional facets of the original acute m-dimensional cone M have a bijection. Notably, each vector (in particular, z^i) representing a generator of the dual cone is the inner normal for some $(m-1)$-dimensional facet of the original cone, and conversely.

Fix a linearly independent subcollection $a^{i1}, \ldots, a^{i(m-1)}$ from $(m-1)$ vectors in the collection $a^1, a^2, \ldots, a^{m+k}$, that generates the $(m-1)$-dimensional facet of the cone M with the vector z^i as its normal. Suppose that relationships (5.33) fail for the subcollection $a^{i1}, \ldots, a^{i(m-1)}$. This means violation of the corresponding inequalities, i.e., there exists index $j \in \{m, \ldots, m+k\}$ such that $\langle a^{ij}, z^i \rangle < 0$. This inequality is inconsistent with the inclusion $z^i \in M^o$. ■

By Lemma 4.1, the vector z^i memorized at Step 4 is a generator of the dual cone M^o, i.e., it is one from the collection $b^1, b^2, \ldots, b^n$.

The next result justifies the algorithm usage for Pareto set reduction.

Theorem 5.8 *Assume that there are given vectors $u^1, u^2, \ldots, u^k$ with at least one positive and at least one negative components that generate a consistent collection of information quanta, i.e., $u^i \succ 0_m$, $i = 1, \ldots, k$,. Then for any set of selectable vectors $C(Y)$ we have*

$$C(Y) \subset \hat{P}(Y) \subset P(Y). \tag{5.35}$$

Here $\hat{P}(Y) = P_g(Y)$ and the vector function

$$g(y) = (\langle b^1, y \rangle, \langle b^2, y \rangle \ldots, \langle b^n, y \rangle), \quad n \geqq m,$$

is constructed from the vectors $b^1, b^2, \ldots, b^n$ yielded by the above algorithm with the vectors $u^1, u^2, \ldots, u^k$ and the m unit vectors of space R^m as the input data.

□ For convenience, denote by $a^1, a^2, \ldots, a^{m+k}$ the vectors $e^1, e^2, \ldots, e^m, u^1, u^2, \ldots, u^k$ mentioned in the hypothesis of this theorem. And let M indicate the cone (without the origin) generated by this collection of vectors, i.e., $M = cone\{a^1, a^2, \ldots, a^{m+k}\} \backslash 0_m$. Since the given collection of information quanta is consistent, this cone is acute and m-dimensional. Consider the dual cone (without the origin)

$$M^o = \{y \in R^m | \langle a^j, y \rangle \geqq 0, \quad j = 1, \ldots, m+k\} \backslash \{0_m\}.$$

It is acute and m-dimensional, too (for details, see [57, 62]).

Under the reasonable choice axioms, the preference relation $\succ$ described in this theorem is a cone relation with an acute convex cone containing the nonnegative orthant R^m_+. Let K be the cone of the relation $\succ$. Therefore, we have the inclusions $R^m \subset M \subset K$, and all the participating cones are acute, convex and m-dimensional. The corresponding sets of nondominated vectors are nested in the inverse order:

$$\mathrm{Ndom}Y \subset \mathrm{Ndom}_M Y \subset P(Y). \tag{5.36}$$

Here $P(Y)$ means the Pareto set, i.e., the set of nondominated vectors with respect to the cone relation with the cone R^m_+; Ndom Y is the set of nondominated vectors with respect to the cone relation $\succ$; $\mathrm{Ndom}_M Y$ stands for the set of non-dominated vectors with respect to the cone relation $\succ_M$ with the cone M, i.e.,

$$\mathrm{Ndom}_M Y = \{y^* \in Y | \text{ there exists no } y \in Y \text{ such that } y - y^* \in M\}.$$

By Lemma 5.4, the above algorithm constructs the generators $b^1, b^2, \ldots, b^n$ of the dual cone M^o. As is well-known [57], the cone M is dual to the cone M^o (the origin should not be taken into account). Hence, the relationship $y - y^* \in M$ is equivalent to the inqualities $\langle b^j, y\rangle \geq \langle b^j, y^*\rangle$, $j = 1, 2, \ldots, n$, and $\mathrm{Ndom}_M Y = \hat{P}(Y)$ in (5.36), where g is the vector function described in the hipothesis of Theorem 5.8. Since for any set of selectable vectors $C(Y)$ we have the inclusion $C(Y) \subset$ Ndom Y, relationships (5.36) imply the desired inclusions (5.35). ■

According to this theorem, the algorithm requires constructing the new vector criterion g, and the Pareto set $\hat{P}(Y)$ in terms of this criterion yields the upper estimate for the unknown set of selectable vectors $C(Y)$ considering the available collection of information quanta. Generally, this estimate is more accurate than the original Pareto set $P(Y)$.

5.3.3 *Example*

Example 5.2 Choose $m = 3$, $k = 2$, $u^1 = (-2,\ 3,\ 1) \succ 0_3$, and $u^2 = (4,\ -1,\ 1) \succ 0_3$. It is easy to verify that the information about the DM's preference relation in the form of these two quanta is consistent. Let us apply the above algorithm to construct the new vector criterion. The input data are the five vectors from the collection $\{e^1,\ e^2,\ e^3,\ u^1,\ u^2\}$, where $e^1 = (1,\ 0,\ 0)$, $e^2 = (0,\ 1,\ 0)$, and $e^3 = (0,\ 0,\ 1)$. The total number of iterations in this algorithm is $C_5^2 = 10$.

Consider the first subcollection $\{e^1,\ e^2\}$. The vector $\{\ e^3\}$ is orthogonal to both of them, and $\langle e^3,\ u^1\rangle = \langle e^3,\ u^2\rangle = 1 > 0$. Hence, the vector $z^1 = e^3$ is memorized.

Take the second subcollection $\{e^1,\ e^3\}$. The vector e^2 is orthogonal to both vectors from this subcollection, but $\langle e^2,\ u^1\rangle = 3 > 0$ and $\langle e^2,\ u^2\rangle = -1 < 0$. This means that the vector e^2 is not memorized.

Next, study the subcollection $\{e^2,\ e^3\}$. Here the orthogonal vector e^1 satisfies the conditions $\langle e^1,\ u^1\rangle = -2 < 0$ and $\langle e^1,\ u^2\rangle = 4 > 0$. And so, this vector is also skipped.

For the subcollection $\{e^1, u^1\}$, an orthogonal vector is, e.g., $(0, 1, -3)$. Since $\langle(0, 1, -3), e^2\rangle = 1 > 0$ and $\langle(0, 1, -3), u^2\rangle = -4 < 0$, we skip this vector too.

For the subcollection $\{e^2, u^1\}$, a possible orthogonal vector is $z^2 = (1, 0, 2)$, which is memorized. In the same fashion, for the subcollections $\{e^3, u^1\}$ and $\{e^1, u^2\}$ we memorize, e.g., the vectors $z^3 = (3, 2, 0)$ and $z^4 = (0, 1, 1)$, respectively. For the subcollection $\{e^2, u^2\}$ the corresponding orthogonal vector is skipped, whereas for the subcollection $\{e^3, u^2\}$ we have to memorize the vector $z^5 = (1, 4, 0)$. And finally, there is nothing to memorize for the subcolletion $\{u^1, u^2\}$.

As a result, we have found the five vectors $z^1, \ldots, z^5$. They correspond to the new vector criterion g with the components $g_1(y) = \langle z^1, y\rangle = y_3$, $g_2(y) = \langle z^2, y\rangle = y_1 + 2y_3$, $g_3(y) = \langle z^3, y\rangle = 3y_1 + 2y_2$, $g_4(y) = \langle z^4, y\rangle = y_2 + y_3$, and $g_5(y) = \langle z^5, y\rangle = y_1 + 4y_2$.

According to Theorem 5.8, the Pareto set in terms of this five-dimensional criterion is a refined upper estimate for the unknown set of selectable vectors.

To obtain a particular solution, choose, e.g., the feasible set $Y = \{y^1, y^2, y^3, y^4\}$ of the form

$$y^1 = (1,\ 4.5,\ 2),\ y^2 = (2,\ 3,\ 1),\ y^3 = (3,\ 2,\ 1.5),\ y^4 = (5,\ 1.5,\ 2).$$

As easily checked, all these vectors are Pareto optimal. Standard calculations show that

$$g(Y) = \{(2, 5, 12, 6.5, 19), (1, 4, 12, 4, 14), (1.5, 6, 13, 3.5, 9), (2, 9, 18, 3.5, 11)\}.$$

The second and third vectors in this set are not Pareto optimal. Consequently, $\hat{P}(Y) = P_g(Y) = \{y^1, y^4\}$, i.e., using the available information we have reduced the Pareto set by 50%.

5.4 Algebraic Algorithm of Vector Criterion Recalculation

5.4.1 Statement of the Problem

Under Axioms 1–4, the preference relation is a cone relation and its convex cone K includes the nonnegative orthant (see Theorem 2.1). If there exists additional information in the form of the relationships $u^k \succ 0_m$, where $u^k \in N^m, k = 1, 2, \ldots, q$, then $u^k \in K, k = 1, 2, \ldots, q$, by the definition of the cone relation $\succ$. Furthermore, all unit vectors of R^m belong to the cone K, i.e., $e^k \in K$, $k = 1, 2, \ldots, m$. Since the cone K is convex, it contains the whole cone generated by the above vectors, i.e.,

$$cone\{e^1, e^2, \ldots, e^m, u^1, u^2, \ldots, u^p\} \backslash \{0_m\} = M \subset K.$$

The chain of inclusions $K \subset M \subset R^m_+$ implies

$$\text{Ndom}\, Y \subset \{y \in R^m |\ \text{there exists no}\ y' \ \text{such that}\ y' - y \in M\} \subset P(Y). \quad (5.37)$$

Consider the cone Q that is dual to the closed cone $M \cup \{0_m\}$. Since the latter is a finitely generated cone, then its dual cone Q can also be represented in the form of a linear nonnegative combination of a finite number of vectors: $Q = cone\{b^1, b^2, \ldots, b^q\}$. In this case, the relationship $y' - y \in M$ is equivalent to the inequalities $\langle b^k, y' - y\rangle \geq 0$ for all $k = 1, 2, \ldots, p$, with at least one inequality here being strict. Indeed, if $y' - y \in M$, then the inequalities hold by the definition of a dual cone. Imagine that all these inequalities hold as equalities. Then the nonzero vector $y' - y$ is orthogonal to the linear hull $L\{b^1, b^2, \ldots, b^q\}$, which implies the inclusion $y' - y \in M$ and also the inclusion $y - y' \in M$ by to the definition of a dual cone. In this case, we simultaneously have the relationships $y' \succ y$ and $y \succ y'$, which are inconsistent with the asymmetric property of the preference relation. The converse seems obvious: the above inequalities imply $y' - y \in M \cup \{0_m\}$ by the definition of a dual cone, but the vector $y' - y$ cannot be zero (there exists at least one strict inequality).

Construct the new vector criterion g in the following way. Choose the scalar product $g_k(y) = \langle b^k, y\rangle, \quad k = 1, 2, \ldots, q$, as its component k. Then the result established above can be rewritten as $y' - y \in M \Leftrightarrow g(y' - y) \geq 0_q$, which is equivalent to $g(y') \geq g(y)$ owing to the linear property of the scalar product. Hence, the set in the right-hand side of (5.37) is none other than the Pareto set in terms of the new vector criterion, i.e.,

$$\begin{aligned} &\{y \in R^m |\ \text{there exists no}\ y' \in R^m \ \text{such that}\ y' - y \in M\} \\ &\quad = \{y \in R^m |\ \text{there exists no}\ y' \in R^m \ \text{such that}\ g(y') \geq g(y)\} = P_g(Y). \end{aligned}$$

These considerations naturally lead to the following statement.

Theorem 5.9 *Assume that there is a given finite collection of vectors $u^k \in N^m$ that satisfy the relationships $u^k \succ 0_m$, $k = 1, 2, \ldots, p$. In addition, let vectors $b^1, b^2, \ldots, b^q$ generate the dual cone for $cone\{u^1, u^2, \ldots, u^p, e^1, e^2, \ldots, e^m\}$. Finally, let the components of the vector criterion $g(y)$ be defined by the equalities $g_k(y) = \langle b^k, y\rangle$, $k = 1, 2, \ldots, q$. Then for any set of selectable vectors $C(Y)$ we have the inclusions*

$$C(Y) \subset P_g(Y) \subset P(Y). \quad (5.38)$$

Therefore, to take into account a finite collection of information quanta, one has to construct the generators of the dual cone for a given acute finitely generated cone (the former problem is reduced to the latter one). This is the subject of the forthcoming subsection.

5.4.2 Construction of Dual Cone Generators

Consider the following problem. Given acute finitely generated $cone\{e^1, e^2, \ldots, e^m, u^1, u^2, \ldots, u^p\}$, it is required to find the generators of its dual acute cone Q.

Generally (i.e., if the cone M has an arbitrary structure), the dual cone may contain no unit vectors of the space, or even may not be acute. In the general case, the problem is solved by the Motzkin-Burger algorithm [7]. However, owing to the special properties of the cone associated with the multicriteria specifics, we can substantially simplify this algorithm.

The dual cone is representable as the intersection of the half-spaces

$$\begin{aligned} Q &= (\cap_{k=1}^{m}\{y \in R^m |\ \langle e^k, y\rangle \geqq 0\}) \cap (\cap_{k=1}^{p}\{y \in R^m |\ \langle u^k, y\rangle \geqq 0\}) \\ &= \{y \in R^m |\ y \geqq 0\} \cap (\cap_{k=1}^{p}\{y \in R^m |\ \langle u^k, y\rangle \geqq 0\}). \end{aligned}$$

Introduce the cones

$$Q_s = \{y \in R^m | y \geqq 0\} \cap \ \cap_{k=1}^{s}\{y \in R^m | \langle u^k, y\rangle \geqq 0\}, s = 0, 1, \ldots, p.$$

Then $Q_{s+1} = Q_s \cap \{y \in R^m |\ \langle u^{s+1}, y\rangle \geqq 0\}$, and the desired cone has the form $Q = Q_p$, while $Q_0 = \cap_{k=1}^{m}\{y \in R^m |\ \langle e^k, y\rangle \geqq 0\} = cone\{e^1, e^2, \ldots, e^m\}$.

Thus, if we know the recalculation algorithm for the vectors generating the intersection of the cone Q_s with the half-space $\{y \in R^m |\ \langle u^{s+1}, y\rangle \geqq 0\}$, then the generators of the cone Q can be obtained in p steps from the generators of the first orthant. This idea is justified by the following statement.

Theorem 5.10 *Divide the indexes of the vectors $a^1, a^2, \ldots, a^r$ into three groups in accordance with the partition of the whole space by the plane $\{y \in R^m |\ \langle u^{s+1}, y\rangle = 0\}$:*

$$A = \{i |\ \langle u^{s+1}, a^i\rangle > 0\},\ B = \{j |\ \langle u^{s+1}, a^j\rangle < 0\},\ C = \{k |\ \langle u^{s+1}, a^k\rangle = 0\}.$$

Then the vectors

$$\begin{aligned} &a^i \ \textit{for all} \ \ i \in A \cup C, \quad d^{ij} = \langle u^{s+1}, a^i\rangle a^j - \langle u^{s+1}, a^j\rangle a^i \\ &\textit{for all} \ \ (i,j) \in A \times B, \end{aligned} \tag{5.39}$$

generate the cone $Q_{s+1} = Q_s \cap \{y \in R^m | \langle u^{s+1}, y\rangle \geqq 0\}$.

☐ It is easy to verify that the vector collection (5.39) belongs to the intersection $Q_s \cap \{y \in R^m | \langle u^{s+1}, y\rangle \geqq 0\}$. Indeed, the vectors d^{ij} are the linear combinations of the vectors $a^1, a^2, \ldots, a^r$ with positive coefficients, and so they do belong to the above intersection. The index sets A and C are organized so that $\langle u^{s+1}, a^i\rangle \geqq 0$ for all $i \in A \cup C$. Finally, $\langle u^{s+1}, d^{ij}\rangle = \langle u^{s+1}, a^i\rangle\langle u^{s+1}, a^j\rangle - \langle u^{s+1}, a^j\rangle\langle u^{s+1}, a^i\rangle = 0$.

It remains to demonstrate that each vector $y \in Q_s$ satisfying the inequality $\langle u^{s+1}, y\rangle \geqq 0$ can be expressed as a linear nonnegative combination of the vectors from collection (5.39). Since $y \in Q_s$, we have the representation $y = \sum_{i=1}^{r} \gamma_i a^i$ where all coefficients γ_i are nonnegative. If $\sum_{i\in A} \gamma_i \langle u^{s+1}, a^i\rangle = 0$, then using the inequality $\langle u^{s+1}, y\rangle \geqq 0$ we get $\sum_{j\in B} \gamma_j \langle u^{s+1}, a^j\rangle = 0$. The sum of nonnegative numbers γ_i with positive (or negative) coefficients is zero if and only if all these numbers are zero. Thereby, the vector $y = \sum_{i\in C} \gamma_i a^i$ has been decomposed with respect to the vectors from collection (5.39).

In the case $\sum_{i\in A} \gamma_i a^i > 0_m$, perform the transformation

$$y = \sum_{i\in A\cup C} \gamma_i a^i + \frac{\sum_{i\in A} \gamma_i \langle u^{s+1}, a^i\rangle}{\sum_{k\in A} \gamma_k \langle u^{s+1}, a^k\rangle} \sum_{j\in B} \gamma_j a^j = \sum_{i\in A\cup C} \gamma_i a^i + \frac{\sum_{i\in A}\sum_{j\in B} \gamma_i\gamma_j \langle u^{s+1}, a^i\rangle a^j}{\sum_{k\in A} \gamma_k \langle u^{s+1}, a^k\rangle}$$

$$= \sum_{i\in A\cup C} \gamma_i a^i + \frac{\sum_{i\in A}\sum_{j\in B} \gamma_i\gamma_j (d^{ij} + \langle u^{s+1}, a^j\rangle a^i)}{\sum_{k\in A} \gamma_k \langle u^{s+1}, a^k\rangle} = \sum_{i\in C} \gamma_i a^i + \sum_{(i,j)\in A\times B} \frac{\gamma_i\gamma_j d^{ij}}{\sum_{k\in A} \gamma_k \langle u^{s+1}, a^k\rangle}$$

$$+ \sum_{i\in A} \gamma_i \left(1 + \frac{\sum_{j\in B} \gamma_j \langle u^{s+1}, a^j\rangle}{\sum_{k\in A} \gamma_k \langle u^{s+1}, a^k\rangle}\right) a^i,$$

which actually represents the vector y as the linear nonnegative combination of vectors (5.39):

$$1 + \frac{\sum_{j\in B} \gamma_j \langle u^{s+1}, a^j\rangle}{\sum_{k\in A} \gamma_k \langle u^{s+1}, a^k\rangle} = \frac{\sum_{k\in A} \gamma_k \langle u^{s+1}, a^k\rangle + \sum_{j\in B} \gamma_j \langle u^{s+1}, a^j\rangle}{\sum_{k\in A} \gamma_k \langle u^{s+1}, a^k\rangle}$$

$$= \frac{\langle u^{s+1}, y\rangle}{\sum_{k\in A} \gamma_k \langle u^{s+1}, a^k\rangle} \geqq 0.$$

■

However, not all vectors of collection (5.39) are the generators the cone Q_{s+1}. To illustrate this, e.g., consider $Q_s = cone\{(3,\ 1,\ 1), (1,\ 3,\ 1), (3,\ 3,\ 1),\ (1,\ 1,\ 1)\}$ and $u^{s+1} = (1,\ -1,\ 0)$. The first vector generating the cone Q_s belongs to the positive half-space, the second to the negative half-space and the rest to the hyperplane $\{y \in R^m |\ \langle u^{s+1}, y\rangle = 0\}$. According to Theorem 5.11, the cone Q_{s+1} is generated by the vector collection $\{(3,\ 1,\ 1), (3,\ 3,\ 1), (1,\ 1,\ 1), (8,\ 8,\ 4)\}$. The last vector clearly is a positive linear combination of the two preceding ones (their doubled sum), and hence it is not a generator.

Theorem 5.11 *A vector a^r is not an edge (generator) in the collection of vectors $a^1, a^2, \ldots, a^r$ generating a certain acute cone if and only if it is a zero vector, or it is collinear to another vector from the collection, or there exist noncollinear vectors $x, y \in cone\{a^1, a^2, \ldots, a^r\}$ and a number $\alpha > 0$ such that $x + y = \alpha \cdot a^r$.*

□ Necessity. If a vector a^r is not an edge, then $a^r \in cone\{a^1, a^2, \ldots, a^r\} = cone\{a^1, a^2, \ldots, a^{r-1}\}$. Hence, the representation $a^r = \sum_{k=1}^{r-1} \gamma_k a^k$ takes place, and also $\gamma_k \geq 0,\ k = 1, 2, \ldots, r-1$. If all coefficients γ_k are zero, then $a^r = 0_m$. If exactly one coefficient is nonzero, then the vector a^r is collinear to the corresponding generator. Finally, assume that at least two coefficients are nonzero. Without loss of generality, let the first of them be positive, i.e., $\gamma_1 > 0$. In this case, $a^r = \gamma_1 a^1 + \sum_{k=2}^{r-1} \gamma_k a^k$. It follows thast the vector a^r is expressed as the sum of two noncollinear vectors or the vectors a^r and a^1 are collinear.

Sufficiency. The inclusion $cone\{a^1, a^2, \ldots, a^{r-1}\} \subset cone\{a^1, a^2, \ldots, a^r\}$ naturally holds. We will show the opposite inclusion. Take a vector $z \in cone\{a^1, a^2, \ldots, a^r\}$. It can be expanded into the sum

$$z = \sum_{k=1}^{r} \gamma_k a^k. \tag{5.40}$$

If a^r is a zero vector, we have $z = \sum_{k=1}^{r-1} \gamma_k a^k \subset cone\{a^1, a^2, \ldots, a^{r-1}\}$. If a^r is collinear to one of the generators, it can be also eliminated from the sum. Consider the case $\alpha \cdot a^r = x + y$, where x and y are two noncollinear vectors from the cone $cone\{a^1, a^2, \ldots, a^r\}$. These vectors admit the representations $x = \sum_{k=1}^{r} \phi_k a^k$ and $y = \sum_{k=1}^{r} \psi_k a^k$. Since the vectors under consideration are noncollinear, there exist $i, j,\ i \neq j$, such that $\phi_i > 0,\ a^i \neq 0_m,\ \psi_j > 0,\ a^j \neq 0_m$. In this case,

$$a^r = \frac{x+y}{\alpha} = \sum_{k=1}^{r-1} \frac{\phi_k + \psi_k}{\alpha} a^k, \quad \left(1 - \frac{\phi_r + \psi_r}{\alpha}\right) a^r = \sum_{k=1}^{r-1} \frac{\phi_k + \psi_k}{\alpha} a^k,$$

and the right-hand side of the last equality contains at least one positive coefficient. The parenthesized expression is not negative (otherwise, $cone\{a^1, a^2, \ldots, a^r\}$ fails to be acute). This expression is not zero, as the right-hand side makes a nonzero vector. Therefore, the parenthesized expression is positive, and the vector a^r is represented in the form a linear nonnegative combination of the generators $a^1, a^2, \ldots, a^{r-1}$. By substituting this representation into sum (5.40), we actually expand the vector z with respect to the generators $a^1, a^2, \ldots, a^{r-1}$. This proves the opposite inclusion. ■

Now, we analyze possible ways to eliminate the vectors that are not edges at each step of the algorithm. Suppose that all vectors generating the cone Q_s are edges. Since $Q_{s+1} \subset Q_s$, all the memorized vectors $a^k, k \in A \cup C$ are still edges. Indeed, if not, we have two possible cases: (1) there exist two noncollinear vectors $x, y \in Q_{s+1} \subset Q_s$ whose sum is collinear to a^k or (2) a certain vector among the ones generating the cone Q_{s+1} is collinear to the vector a^k. Case (1) directly contradicts the fact that a^k is an edge in the cone Q_s. In case (2), this collinear vector is d^{ij}, as the vectors generating Q_s represent edges. As a result, the vector $a^k = \langle u^{s+1}, a^i\rangle a^j + \langle -u^{s+1}, a^j\rangle a^i$ is rewritten as the sum of two noncollinear vectors from Q_s, which means that it is not an edge.

As is well-known, a generator (edge) of a cone is a normal to an $(m-1)$-dimensional facet of the dual cone. This property leads to the following result.

Theorem 5.12 *A vector d^{ij} of the cone Q_{s+1} that is not collinear to any other vector $d^{i'j'}$ forms a generator if and only if $rang T(d^{ij}) = m - 1$, where*

$$T(d^{ij}) = (\bigcup_{k=1}^{m} \{e^k | \langle d^{ij}, e^k\rangle = 0\}) \cup (\bigcup_{k=1}^{s+1} \{e^k | \langle d^{ij}, e^k\rangle = 0\}).$$

□ Necessity. The rank of the vector collection $T(d^{ij})$ is not m, since an edge is not a zero vector. Assume that this rank is smaller than $m-1$. Then the linear hull of the vectors $T(d^{ij}) \cup \{d^{ij}\}$ has a smaller dimension than space R^m, and hence there exists a nonzero vector z orthogonal to d^{ij} and also to all vectors from $T(d^{ij})$.

By the construction procedure of the cone Q_{s+1}, the vector d^{ij} satisfies the inequalities

$$\langle d^{ij}, e^k\rangle \geqq 0, k = 1, 2, \ldots, m, \quad \langle d^{ij}, u^k\rangle \geqq 0, k = 1, 2, \ldots, s+1$$

Choose a number $\varepsilon > 0$ so that

$$\begin{aligned} 0 \leqq \varepsilon |\langle z, e^k\rangle| \leqq \langle d^{ij}, e^k\rangle, \quad k = 1, 2, \ldots, m, \\ 0 \leqq \varepsilon |\langle z, u^k\rangle| \leqq \langle d^{ij}, u^k\rangle, \quad k = 1, 2, \ldots, s+1. \end{aligned} \tag{5.41}$$

This is possible, since if the right-hand size is $\langle d^{ij}, e^k\rangle = 0$, we have $e^k \in T(d^{ij})$ and $\langle z, e^k\rangle = 0$; if the right-hand side forms a positive number, the number ε can be chosen sufficiently small.

The conditions imposed on ε allow stating that

$$\begin{aligned} \langle d^{ij} \pm \varepsilon z, e^k\rangle \geqq 0, \quad k = 1, 2, \ldots, m, \\ \langle d^{ij} \pm \varepsilon u^k, e^k\rangle \geqq 0, \quad k = 1, 2, \ldots, s+1. \end{aligned}$$

Therefore, the vectors $d^{ij} \pm \varepsilon z$ belong to the cone Q_{s+1}. And they are not collinear, since $d^{ij} \perp z$. This means that the vector d^{ij} can be expressed as the half-sum of two noncollinear vectors from Q_{s+1}, and it is not an edge accordingly.

Sufficiency. Now, suppose that (despite the equality $rangT(d^{ij}) = m - 1$) the vector d^{ij} is not an edge. Note that d^{ij} is a nonzero vector (otherwise, all unit vectors of space R^m would enter $T(d^{ij})$, yielding rank m for the resulting collection). It has been shown above that the vectors a^k, $k \in A \cup C$, cannot be collinear to d^{ij}. Assume that there exist vectors $x, y \in Q_{s+1}$ such that $x + y = \alpha \cdot d^{ij}$ for some $\alpha > 0$. Take an arbitrary vector $z \in T(d^{ij})$. The inequalities $\langle z, x \rangle \geqq 0$, $\langle z, y \rangle \geqq 0$ hold due to $x, y \in Q_{s+1}$. However, it follows from the inclusion $z \in T(d^{ij})$ that $\langle z, x + y \rangle = \langle z, d^{ij} \rangle = 0$. The sum of two nonnegative numbers can be zero only if they are both zero: $\langle z, x \rangle = \langle z, y \rangle = 0$. And so, both vectors x and y are orthogonal to all vectors from $T(d^{ij})$. But in this case they must be collinear, since $rangT(d^{ij}) = m - 1$. This contradiction completes the proof of Theorem 5.12. ■

Rank calculation can be replaced with the comparison of vector collections for different generators. Then we have the following criterion to eliminate the vectors that are not generators.

Theorem 5.13 *A nonzero vector d^{ij} of the cone Q_{s+1} is not a generator if and only if there exists a vector q (a^k or $d^{i'j'}$) of this cone such that $T(d^{ij}) \subset T(q)$.*

□ Necessity. If the vector d^{ij} is not an edge, then two situations are possible, namely, (1) there exists a collinear vector $d^{i'j'}$ or (2) $rangT(d^{ij}) < m - 1$. In situation (1), we have $T(d^{ij}) \subset T(d^{i'j'})$ and the necessity is proved accordingly. In situation (2), there exist a vector z being orthogonal to d^{ij} and to all vectors from $T(d^{ij})$ and a number $\varepsilon > 0$ satisfying inequalities (5.41). Choose the maximum number among all such numbers ε, i.e.,

$$\varepsilon = \min_{t \in T \setminus T(d^{ij})} \frac{\langle d^{ij}, t \rangle}{|\langle z, t \rangle|}, \tag{5.42}$$

where $T = \{u^1, u^2, \ldots, u^{s+1}, e^1, e^2, \ldots, e^m\}$. By definition, both vectors $d^{ij} \pm \varepsilon z$ belong to the cone Q_{s+1}, and one of them is also orthogonal to the vector $\bar{t} \in T(d^{ij})$ that corresponds to the minimum in (5.42). Let us expand this vector with respect to the edges of the cone Q_{s+1} except the vector d^{ij}. Since $d^{ij} \pm \varepsilon z \in Q_{s+1}$, there exists at least one vector q entering this expansion with a positive coefficient. As it belongs to Q_{s+1}, the inequality $\langle q, t \rangle \geqq 0$ holds for all $t \in T$. But the sum of such vectors with positive coefficients yields the vector that is orthogonal to $\{\bar{t}\} \cup T(d^{ij})$. This is possible only if $\{\bar{t}\} \cup T(d^{ij}) \subset T(q)$, and hence $T(d^{ij}) \subset T(q)$.

Sufficiency. If the inclusion $T(d^{ij}) \subset T(q)$ takes place, then $rankT(d^{ij}) \leqq m - 1$. Under the equality $rankT(d^{ij}) = m - 1$, the vectors d^{ij} and q belong to the one-dimensional linear space, thereby being collinear. This means that d^{ij} is not an edge. If $rankT(d^{ij}) < m - 1$, then the desired result follows directly from Theorem 5.12. ■

Note that all the vectors d^{ij} are nonzero, since they represent nonnegative linear combinations of the nonzero edges a^i, a^j.

Finally, consider another special case that may occur in the algorithm. Let $A = \varnothing$ at some step during the construction of the cone Q_{s+1}. That is, all vectors q of the cone Q_s satisfy the inequality $\langle q, u^{s+1} \rangle \leqq 0$. But then $\langle q, -u^{s+1} \rangle \geqq 0$, and the vector $-u^{s+1}$ belongs to the cone that is dual to Q_s, i.e., $-u^{s+1} \in cone\{u^1, u^2, \ldots, u^s, e^1, e^2, \ldots, e^m\}$. The compatibility axiom dictates that $e^k \succ 0_m$, $k = 1, 2, \ldots, m$, while the definition of an information quantum leads to the relationships $u^k \succ 0_m$, $k = 1, 2, \ldots, s$. Owing to the invariance axiom, we deduce that $-u^{s+1} \succ 0_m$. At the same time, the relationship $u^{s+1} \succ 0_m$ holds by the above specification of an information quantum. The resulting contradiction $u^{s+1} + (-u^{s+1}) = 0_m \succ 0_m$ indicates that the available collection of information quanta is inconsistent.

5.4.3 *Algorithm of Information Quanta Consideration*

Now, we describe the algorithm for taking into account the available information in the form of an arbitrary collection of vectors $u^1, u^2, \ldots, u^p \in N^m$.

At the start of the algorithm, it is necessary to initialize *the list of generators* by the unit vectors $e^1, e^2, \ldots, e^m$.

At each step $s = 1, 2, \ldots, p$ we have the list of the generators $q^1, q^2, \ldots, q^{r_s}$ divided into three groups as follows:

$A = \{i | \langle u^s, q^i \rangle > 0\}$, $B = \{j | \langle u^s, q^j \rangle < 0\}$, $C = \{k | \langle u^s, q^k \rangle = 0\}$.

If $A = \varnothing$, the collection of the existing information quanta that includes the vectors $u^1, u^2, \ldots, u^p$ is inconsistent and the algorithm halts. Otherwise, the vectors q^i for $i \in A \cup C$ are added to the new list of vectors. Next, iterating over all pairs $(i,j) \in A \times B$, it is necessary to add the vector $d^{ij} = \langle u^s, a^i \rangle a^j - \langle u^s, a^j \rangle a^i$ to the new list if the set

$$T(d^{ij}) = (\bigcup_{k=1}^{m} \{e^k | \langle d^{ij}, e^k \rangle = 0\}) \cup (\bigcup_{k=1}^{s} \{u^k | \langle d^{ij}, u^k \rangle = 0\})$$

is not contained in any similar set $T(q)$ for the generatrix q from the new list.

The vectors $q^1, q^2, \ldots, q^{r_p}$ included in the list after p steps are the generators of the dual cone. Based on these vectors, we may construct the new vector criterion using the formula $g(y) = (g_1(y), g_2(y), \ldots, g_{r_p}(y))$, where $g_k(y) = \langle q^k, y \rangle$, $k = 1, 2, \ldots, r_p$. And the Pareto set in terms of the new vector criterion g yields the desired upper estimate (5.38).

5.5 Reduction of Finite Pareto Set

5.5.1 Main Idea

Consider situation where the available information about the DM's preference relation contains an arbitrary finite collection of vector pairs

$$u^i, v^i \in R^m, \quad u^i - v^i \in N^m \cap K, \quad i = 1, 2, \ldots, k,$$

satisfying the relationships $u^i \succ v^i$, $i = 1, 2, \ldots, k$, where K denotes the acute convex cone of the cone preference relation $\succ$. Recall that the set N^m is formed by all m-dimensional vectors having at least one positive and at least one negative components.

Introduce the convex cone M (without the origin) generated by the vectors

$$e^1, e^2, \ldots, e^m, u^1 - v^1, \ldots, u^k - v^k. \tag{5.43}$$

We have the inclusion $M \subset K$ and hence the cone M is acute. Moreover, $R^m_+ \subset M$.

Denote by $\succ_M$ the cone relation with the cone M. This relation represents the same class of relations as the original relation $\succ$. For justice' sake, note that the relation $\succ$ also satisfies Axiom 1, whereas the relation $\succ_M$ may not. But this aspect is not crucial for further exposition.

Consequently, we have two cone relations, $\succ$ and $\succ_M$, that are interconnected via the implication

$$y' \succ_M y'' \quad \Rightarrow \quad y' \succ y''$$

for all $y', y'' \in R^m$. This interconnection holds due to the inclusion $M \subset K$, yielding

$$\text{Ndom}\, Y \subset \text{Ndom}_M\, Y, \tag{5.44}$$

where

$\text{Ndom}_M Y = \{y^* \in Y |$ there exists no $y \in Y$ such that $y \succ_M y^*\}$.

Inclusion (5.44) means that the set $\text{Ndom}_M Y$ is an upper estimate for the set of nondominated vectors $\text{Ndom} Y$, and so for any set of selectable vectors $C(Y)$. Since the set Y is finite, the direct enumeration of all pairs of its elements allows for obtaining the set $\text{Ndom}_M Y$. The latter is generally narrower than the Pareto set, and we actually reduce the Pareto set by eliminating some Pareto optimal vectors. This makes the key idea of the approach suggested below.

5.5.2 Majorant Relation

The cone relation $\succ_M$ with the acute convex cone M (without the origin) generated by vectors (5.43) is called *the majorant relation*. This relation is used below to construct an upper estimate (majorant) for the set of selectable vectors, which explains its name.

The suggested approach proceeds from the following statement.

Theorem 5.14 *Let $y', y'' \in R^m$, $y' \neq y''$. The relationship $y' \succ_M y''$ holds if and only if the canonical linear programming problem*

$$\xi_1 + \xi_2 + \ldots + \xi_m \to \min$$

subject to the constraints

$$\sum_{i=1}^{m} \lambda_i e_s^i \operatorname{sign}(y_s' - y_s'') + \sum_{i=1}^{k} \mu_i (u_s^i - v_s^i) \operatorname{sign}(y_s' - y_s'') + \xi_s = |y_s' - y_s''|, \quad s = 1, 2, \ldots, m, \tag{5.45}$$

and

$$\lambda_1, \lambda_2, \ldots, \lambda_m, \mu_1, \mu_2, \ldots, \mu_k, \xi_1, \xi_2, \ldots, \xi_m \geqq 0,$$

has the zero optimal value of the goal function.

Here

$$\operatorname{sign}(a) = \begin{cases} 1, & \text{if } a > 0 \text{ or } a = 0, \\ -1, & \text{if } a < 0. \end{cases}$$

□ First of all, take notice that the relationship $y' \succ_M y''$ holds or not simultaneously with the inclusion $y' - y'' \in M$, which is equivalent to the equality

$$\sum_{i=1}^{m} \lambda_i e^i + \sum_{i=1}^{k} \mu_i (u^i - v^i) = y' - y'' \tag{5.46}$$

for some N-solution $\lambda_1, \lambda_2, \ldots, \lambda_m, \mu_1, \mu_2, \ldots, \mu_k$. By-turn, the above equality (5.46) is true if and only if

$$\sum_{i=1}^{m} \lambda_i e_s^i \operatorname{sign}(y_s' - y_s'') + \sum_{i=1}^{k} \mu_i (u_s^i - v_s^i) \operatorname{sign}(y_s' - y_s'') = |y_s' - y_s''|, \quad s = 1, 2, \ldots, m. \tag{5.47}$$

Next, for equalities (5.47) to be true for some N-solution $\lambda_1, \lambda_2, \ldots, \lambda_m, \mu_1, \mu_2, \ldots, \mu_k$, a necessary and sufficient condition is that the canonical linear programming problem (5.45) has the optimal solution with $\xi_1 = \xi_2 = \ldots = \xi_m = 0$. The last requirement is equivalent to the zero optimal value of the goal function in the linear programming problem (5.45). ■

In accordance with Theorem 5.14, the relationship $y' \succ_M y''$ is verified by solving the linear programming problem (5.45). This can be done by the well-known simplex method. Such a verification procedure of the relationship $y' \succ_M y''$ seems convenient for the general upper estimation algorithm with a finite set of feasible vectors Y. Sometimes, it is required to solve a low-dimensional problem "manually"; then one should use the following result that represents a special case of Theorem 5.14 established in its proof.

Corollary 5.3 *Let $y', y'' \in R^m$, $y' \neq y''$. The relationship $y' \succ_M y''$ holds if and only if the system of inhomogeneous linear equations* (5.46) *has the N-solution* $\lambda_1, \lambda_2, \ldots, \lambda_m, \mu_1, \mu_2, \ldots, \mu_k$.

5.5.3 *Example*

Let $m = 3, \;\; k = 2, \;\; Y = \{y^1, y^2, y^3, y^4\}$, where

$y^1 = (1, 4.5, 2)$, $y^2 = (2, 3, 1)$, $y^3 = (3, 2, 1.5)$, $y^4 = (5, 1.5, 2)$,

$u^1 = (0, 5, 1)$, $v^1 = (2, 2, 0)$, $u^2 = (5, 0, 2)$, $v^2 = (1, 1, 1)$.

Since

$$u^1 - v^1 = (-2, 3, 1) \succ 0_3, \quad u^2 - v^2 = (4, -1, 1) \succ 0_3,$$

two pairs of vectors u^1, v^1 and u^2, v^2 specify a pair of information quanta. The first quantum states that the group of criteria f_2 and f_3 is more important than criterion f_1. According to the second quantum, the group composed of criteria f_1 and f_3 has higher importance than criterion f_2.

First, verify the consistency of these two quanta using Theorem 4.2. Write the corresponding system of homogeneous linear equations (4.4):

$$\begin{aligned} &\lambda_1 - 2\mu_1 + 4\mu_2 = 0, \\ &\lambda_2 + 3\mu_1 - \mu_2 = 0, \\ &\lambda_3 + \mu_1 + \mu_2 = 0. \end{aligned}$$

The last equation implies $\lambda_3 = \mu_1 = \mu_2 = 0$, since the numbers λ_3, μ_1, μ_2 are nonnegative. Then we obtain $\lambda_1 = \lambda_2 = 0$ from the first and second equations. Hence, the system of linear equations has no N-solutions and, by Theorem 4.2, the two pairs of vectors are consistent.

Now, construct an upper estimate for the set of nondominated vectors Ndom$Y$ (ergo, for the set of selectable vectors $C(Y)$). To this end, write the system of linear equations (5.46) for the vectors $y' = y^1$ and $y'' = y^2$:

$$\begin{aligned} \lambda_1 - 2\mu_1 + 4\mu_2 &= -1, \\ \lambda_2 + 3\mu_1 - \mu_2 &= 1.5, \\ \lambda_3 + \mu_1 + \mu_2 &= 1. \end{aligned}$$

It has the N-solution $\lambda_1 = \lambda_2 = \mu_2 = 0$, $\lambda_3 = \mu_1 = 0.5$. The relationship $y^1 \succ_M y^2$ holds accordingly and the vector y^2 does not enter the set of nondominated vectors $\mathrm{Ndom}_M Y$.

For the vectors $y' = y^4$ and $y'' = y^3$, the system of linear equations (5.46) acquires the form

$$\begin{aligned} \lambda_1 - 2\mu_1 + 4\mu_2 &= 2, \\ \lambda_2 + 3\mu_1 - \mu_2 &= -0.5, \\ \lambda_3 + \mu_1 + \mu_2 &= 0.5. \end{aligned}$$

This system has the N-solution $\lambda_1 = \lambda_2 = \lambda_3 = \mu_1 = 0$, $\mu_2 = 0.5$; and so, the vector y^3 does not belong to the set of nondominated vectors $\mathrm{Ndom}_M Y$ too.

For the vectors $y' = y^1$, $y'' = y^4$ and $y' = y^4$, $y'' = y^1$, the system of linear equations (5.46) is given by

$$\begin{aligned} \lambda_1 - 2\mu_1 + 4\mu_2 &= -4, \\ \lambda_2 + 3\mu_1 - \mu_2 &= 3, \\ \lambda_3 + \mu_1 + \mu_2 &= 0, \end{aligned} \quad \text{and} \quad \begin{aligned} \lambda_1 - 2\mu_1 + 4\mu_2 &= 4, \\ \lambda_2 + 3\mu_1 - \mu_2 &= -3, \\ \lambda_3 + \mu_1 + \mu_2 &= 0. \end{aligned}$$

None of these systems has N-solutions, and the relationships $y^1 \succ_M y^4$, $y^4 \succ_M y^1$ both fail.

We have obtained the two-element set of nondominated vectors

$$\mathrm{Ndom}_M Y = \{y^1, y^4\}.$$

This set yields an upper estimate for arbitrary set of selectable vectors $C(Y)$, i.e., $C(Y) \subset \{y^1, y^4\}$. Clearly, none of the vectors y^2, y^3 can be selected.

5.5.4 Upper Estimation Algorithm

Assume that the vector set Y is finite, i.e.,

$$Y = \{y^1, y^2, \ldots, y^N\}.$$

The design algorithm for the set of nondominated vectors $\mathrm{Ndom}_M Y$ consists of eight steps.

Step 1 Verify the consistency of the collection of vector pairs $u^i, v^i \in R^m$ that satisfy $u^i - v^i \in N^m$, $i = 1, 2, \ldots, k$. This verification is reduced to the solution of the canonical linear programming problem (4.6). If the optimal value of the goal function is zero, terminate the calculations (the given collection of vector pairs is inconsistent). Otherwise, move to the next step.

Step 2 Assign $i = 1$, $j = 2$, $\mathrm{Ndom}_M Y = Y$, thereby forming the so-called *current set of nondominated vectors* (this set yields the desired upper estimate at the end of the algorithm). The algorithm is organized so that this estimate is obtained from Y by the sequential elimination of dominated vectors.

Step 3 Verify the relationship $y^i \succ_M y^j$ by solving the linear programming problem (5.45) for $y' = y^i$, $y'' = y^j$. If the optimal value of the goal function is zero, move to Step 4. Otherwise, move to Step 6.

Step 4 Eliminate the vector y^j from the current set of nondominated vectors $\mathrm{Ndom}_M Y$, since the former might not be contained in the latter.

Step 5 Verify the inequality $j < N$. If it holds, set $j = j + 1$ and get back to Step 3. Otherwise, move to Step 8.

Step 6 Verify the relationship $y^j \succ_M y^i$ by solving the linear programming problem (5.45) for $y' = y^j$, $y'' = y^i$. If the optimal value of the goal function is zero, move to Step 7. Otherwise, get back to Step 5.

Step 7 Eliminate the vector y^i from the current set of nondominated vectors $\mathrm{Ndom}_M Y$.

Step 8 Verify the inequality $i < N - 1$. If it holds, set $i = i + 1$, then $j = i + 1$ and get back to Step 3. Otherwise (which means that $i \geqq N - 1$), finish the calculations. The set of nondominated vectors $\mathrm{Ndom}_M Y$ is constructed.

Chapter 6
Completeness Property of Information Quanta

In this chapter, we justify theoretically the original axiomatic approach to Pareto set reduction based on a finite collection of information quanta. Here the exposition seems most difficult in mathematical terms, but the readers with an insufficient background may skip it without losing the comprehension of further material.

The whole essence of the results derived below can be expressed as follows. Information in the form of quanta is complete: for any multicriteria choice problem from a definite (rather wide) class, it is possible to find the unknown set of non-dominated vectors (nondominated alternatives) with an arbitrary accuracy only based on such information. Moreover, if the number of feasible vectors is finite, then the set of nondominated vectors can be constructed precisely. In other words, by eliciting information quanta about the DM's preference relation, one may successfully construct the set of nondominated alternatives (vectors) without involving other types of information.

6.1 Preliminary Analysis

6.1.1 Problem Statement

An available information quantum allows for eliminating certain Pareto optimal vectors as unacceptable for sure. This gives a refined upper estimate (approximation) for the set of selectable vectors in comparison with the Pareto set. If there exists a finite collection of such information quanta, then hopefully it can be used to construct even a better (more precise) upper estimate, i.e., a narrower set. According to general considerations, a larger collection of information quanta yields a more precise upper estimate. And the following question arises immediately: what are the applicability limits for a finite collection of information quanta about the DM's preference relation?

© Springer International Publishing AG 2018
V.D. Noghin, *Reduction of the Pareto Set*, Studies in Systems, Decision and Control 126, https://doi.org/10.1007/978-3-319-67873-3_6

Note an important aspect prior to further analysis. Owing to Lemma 1.2, the set of selectable vectors is contained in the set of nondominated vectors. Clearly, any subset of the set of nondominated vectors can be selected in the multicriteria choice problems satisfying Axioms 1–4. This means that information about the DM's preference relation and an existing collection of criteria satisfying Axioms 1–4 do not assist in eliminating any nondominated vector as unacceptable for sure. And the set of nondominated vectors is the narrowest upper estimate for the set of selectable vectors in the model under consideration. Therefore, in the sequel we consider the approximation of the set of nondominated vectors instead of the set of selectable vectors.

The above question can be further specified in the following way. *Is it possible to obtain an arbitrary precise representation for the unknown set of nondominated vectors using only a finite collection of information quanta*? In principle, the answer to this question is affirmative, as it will be illustrated below. The reservation "in principle" indicates that it is necessary to restrict slightly the class of the multicriteria choice problems satisfying Axioms 1–4.

For a definite class of multicriteria choice problems, all we need is to extract and utilize information quanta, see the details below. It is quite enough to obtain (at least, theoretically) the unknown set of nondominated vectors with an arbitrary accuracy. This situation testifies to the crucial role of information quanta for decision-making in multicriteria environment.

6.1.2 Geometrical Aspects

Let us formulate the above question in geometrical terms.

According to Definition 3.3, the presence of an information quantum means that there is a given vector $u \in N^m$ with at least one positive and at least one negative components satisfying the relationship $u \succ 0_m$. For a finite collection of such information quanta, we accordingly have a collection of vectors $u^i \in N^m$ satisfying the relationships $u^i \succ 0_m$, $i = 1, 2, \ldots, k$. If this collection of vectors (more specifically, the collection of vector pairs $u^i, 0_m$, $i = 1, 2, \ldots, k$) is consistent, then the convex cone M generated by the vectors $e^1, e^2, \ldots, e^m, u^1, u^2, \ldots, u^k$ represents the aggregate of all N-combinations of these vectors and also an acute convex cone (without the origin). It defines a cone relation further denoted by $\succ_M$.

The question about the completeness of information quanta collections (see the previous subsection) can be easily restated in geometrical terms. How close to the unknown preference relation $\succ$ can be the relation $\succ_M$ obtained using different finite consistent collections of vectors $u^1, u^2, \ldots, u^k$ only? Equivalently, *is it possible in principle to make the relation $\succ_M$ arbitrarily close to the unknown preference relation $\succ$ by choosing the above collection of vectors*?

To simplify further solution, let us translate this problem into the plane of cones: *is it possible to obtain a cone M that is arbitrarily close to the unknown cone* K^1 *by choosing the collection of vectors* $u^1, u^2, \ldots, u^k$? Note that k is not fixed and can be an arbitrary finite number.

The cone K is an arbitrary acute convex cone without the origin. The cone M belongs to the same class as K, being acute, convex and without the origin. In contrast to K, the cone M is generated by a finite number of vectors (a finitely generated cone) and hence polyhedral. In such statement, the completeness problem of information quanta has much in common with a standard convex analysis problem where an arbitrary convex compact set is approximated by a polyhedron. As is well-known, this approximation problem has a solution, i.e., an arbitrary convex closed bounded set can be approximated by a polyhedron with an arbitrary accuracy. And we have every reason to expect that the same problem for cones (approximate an arbitrary convex cone by a polyhedral cone) would be solvable. But first of all it is necessary to agree about the distance between convex cones.

6.1.3 Distance Between Cones

Let A and B be arbitrary non-empty convex subsets of space R^m. *The Hausdorff distance* [23] between these sets is defined by the formula

$$\operatorname{dist}(A, B) = \inf\{r \in R_+ | A \subset (B)_r, B \subset (A)_r\},$$

where R_+ denotes the set of positive real numbers,

$$(A)_r = \bigcup_{y \in A} U_r(y), \ (B)_r = \bigcup_{y \in B} U_r(y),$$

and $U_r(y)$ $(r > 0)$ is a closed ball in R^m having radius r and center at the point y, i.e.,

$$U_r(y) = \{z \in R^m | \, \|z - y\| \leqq r\}.$$

Here $\|a\|$ designates the Euclidean norm (length) of a vector $a \in R^m$, that is,

$$\|a\| = \sqrt{a_1^2 + a_2^2 + \ldots + a_m^2}.$$

In a special case where A and B are singletons $\{a\}$ and $\{b\}$, respectively, the Hausdorff distance between them coincides with the Euclidean distance, being the norm of the difference between the corresponding vectors, $\|a - b\|$.

[1]Recall that K is an acute convex cone of the preference relation $\succ$.

The next result shows that the direct application of the Hausdorff distance to measure the distance between two convex cones causes some difficulties (they will be eliminated, though).

Lemma 6.1 *Let K_1 and K_2 be two arbitrary convex cones in space R^m that do not contain the origin and $\overline{K}_1 \neq \overline{K}_2$, where overline means the closure of a set.*[2] *Then*

$$\mathrm{dist}(K_1, K_2) = +\infty.$$

□ According to the inequality $\overline{K}_1 \neq \overline{K}_2$, there exists a point $y \in R^m$ such that $y \in \overline{K}_1, y \notin \overline{K}_2$, or there exists a point $y \in R^m$ such that $y \in \overline{K}_2, y \notin \overline{K}_1$. For definiteness, consider the first case only, as the second case can be studied by analogy.

The relationships $y \in \bar{K}_1, y \notin \bar{K}_2$ lead to the existence of a point y' such that $y' \neq 0_m$, $y' \in K_1, y' \notin \overline{K}_2$. Consider a ray that comes from the origin and passes through y', as a special case of a cone. Denote this ray by l. Obviously, it satisfies the relationships $l \subset K_1, l \not\subset \overline{K}_2$.

The norm $||y' - y||$ is a continuous function of the variables $y_1, y_2, \ldots, y_m$ and bounded below on the cone K_2. Hence, there exists a limit point $\hat{y} \in R^m$ in the set K_2 such that

$$\inf_{y \in K_2} ||y' - y|| = ||y' - \hat{y}||$$

and $\hat{y} \neq y'$. Construct a sequence $\{y^k\}_{k=1}^{\infty}$ of points on the ray l by choosing

$$y^k = ky', \; k = 1, 2, \ldots,$$

For this sequence, we have

$$\inf_{y \in K_2} \left\|y^k - y\right\| = \inf_{y \in K_2} ||ky' - y|| = k \inf_{y \in K_2} \left\|y' - \frac{y}{k}\right\| = k \inf_{ky \in K_2} ||y' - y|| = k||y' - \hat{y}||_{k \to +\infty} \\ \to +\infty,$$

which immediately gives the required result dist $(K_1, K_2) = +\infty$. ■

This lemma states that the Hausdorff distance between two "essentially noncoincident" cones (whose closures do not coincide) is $+\infty$. And so, this distance does not represent an appropriate closeness measure for the cones of binary relations.

Let K be a convex cone in space R^m and Y be a subset of this space. Introduce the set

$$Y_z = \{y \in Y | \; y - z \in K\}$$

for each $z \in Y$. If there exists a positive constant r such that the inequality

---

[2] *The closure of a set* contains this set together with all its limit points.

$$\sup_{y \in Y_z} \|y - z\| \leqq r$$

holds for any $z \in Y$, then the set Y is called *K-bounded.* As easily checked, each bounded set is K-bounded, but the converse generally fails.

Now, let Y be the set of feasible vectors and K be the convex cone of the cone preference relation $\succ$. Assume that the relationship $y' \succ y''$ takes place for vectors $y', y'' \in R^m$. This is equivalent to the inclusion $y' - y'' \in K$. If the set Y is also K-bounded, then we obtain the inequality $\|y' - y''\| \leqq r$, which is equivalent to the inclusion

$$y' - y'' \in K \cap U_r(0_m).$$

Therefore, for the K-bounded set of feasible vectors Y,

$$y' - y'' \in K \Leftrightarrow y' - y'' \in K \cap U_r(0_m).$$

In other words, for the K-bounded set Y, the closeness of the cones is equivalent to the closeness of their parts located in the ball $U_r(0_m)$.

The above considerations naturally bring to the following definition. *The distance between cones K_1 and K_2*, further denoted by $d_r(K_1, K_2)$, is

$$d_r(K_1, K_2) = \text{dist}(K_1 \cap U_r(0_m),\ K_2 \cap U_r(0_m)), \tag{6.1}$$

where r means a sufficiently large positive number. This distance has the standard properties of a metric, namely, for any convex cones K_1, K_2, and K_3,

(1) $d_r(K_1, K_2) \geqq 0$,
(2) $d_r(K_1, K_2) = 0 \Leftrightarrow \bar{K}_1 = \bar{K}_2$,
(3) $d_r(K_1, K_2) = d_r(K_2, K_1)$
(4) $d_r(K_1, K_3) \leqq d_r(K_1, K_2) + d_r(K_2, K_3)$.

Let $K_3 \subset K_2 \subset K_1$. It is easy to verify that

$$d_r(K_1, K_2) \leqq d_r(K_1, K_3).$$

6.2 First Completeness Theorem

6.2.1 *Statement of Mathematical Problem*

By virtue of Axioms 2–4, the binary preference relation $\succ$ guiding the DM's behavior is a cone relation with an acute convex cone K without the origin. And so, consider an arbitrary acute convex cone K, $K \subset R^m$, that does not contain the origin and includes the nonnegative orthant R^m_+ due to the Pareto axiom. Note that generally K is not a polyhedral cone.

As established in Sect. 6.1, the existence of a finite collection of information quanta is equivalent to the specification of a certain consistent finite collection of vectors $u^1, u^2, \ldots, u^k \in N^m$, that together with the unit vectors $e^1, e^2, \ldots, e^m$ generate a polyhedral cone M contained in the cone K.

The question can be stated in mathematical terms as follows: *is it possible to make the distance $d_r(K, M)$ between the cones K and M arbitrarily small by choosing the vectors $u^1, u^2, \ldots, u^k$ (where k is a finite unfixed number)*?

The next theorem gives the answer.

6.2.2 First Completeness Theorem

Теорема 6.1 *Let K be an arbitrary acute convex cone without the origin, and also $K \subset R^m$, $K \supset R^m_+$, $K \neq R^m_+$. Fix an arbitrary positive number r. Then for any positive number ε there exists a finite collection of vectors*

$$\{u^i\}_{i=1}^k \subset R^m, \quad u^i \in N^m \cap K \cap U_r(0_m), \quad i = 1, 2, \ldots, k,$$

such that

$$d_r(K, \text{cone}\{e^1, e^2, \ldots, e^m, u^1, u^2, \ldots, u^k\}) < \varepsilon, \tag{6.2}$$

where cone$\{e^1, e^2, \ldots, e^m, u^1, u^2, \ldots, u^k\}$ *is the convex cone generated by the finite collection of vectors $e^1, e^2, \ldots, e^m, u^1, u^2, \ldots, u^k$.*

Moreover, the components of all vectors u^i can be assigned rational numbers.

□ Denote

$$\hat{K} = K \cap U_r(0_m), \quad \text{int}\,\hat{K} = \text{int}\,K \cap \text{int}\,U_r(0_m).$$

Fix an arbitrary positive number ε. Introduce m-dimensional regular grid with uniform spacing $\varepsilon/2\sqrt{m}$. Obviously, the hypercube of this grid has diagonal's length $\varepsilon/2$.

Extract all cubes of the grid that intersect with the set int$\hat{K}$. Since the latter is a bounded set, the number of such cubes appears finite. Denote them by $\Pi_1, \Pi_2, \ldots, \Pi_l$ and let

$$\Pi = \bigcup_{j=1}^{l} \Pi_j.$$

By the design procedure, $\operatorname{int}\hat{K} \subset \Pi$. Hence, the closure of the set $\operatorname{int}\hat{K}$ is a subset of the closure of the set Π that coincides with Π due to its closedness. At the same time, $\hat{K}$ is contained in the closure of the set $\operatorname{int}\hat{K}$ and therefore $\hat{K} \subset \Pi$. Thus, Π covers the set $\hat{K}$.

In each intersection $\Pi_j \cap \operatorname{int}\hat{K}$ there exists a point u^j with rational components, $j = 1, 2, \ldots, l$. Let P be the convex hull[3] of all such points u^j. This set is a certain polyhedron.

As $\hat{K}$ is a convex set and $u^j \in \hat{K}$, $j = 1, 2, \ldots, l$, then $P \subset \hat{K}$ and hence

$$P \subset (\hat{K})_\varepsilon. \tag{6.3}$$

On the other hand, there exist points that belong to $\hat{K}$ and do not belong to P simultaneously. Since Π covers $\hat{K}$, the distance between each such point and P does not exceed $\varepsilon/2$. We accordingly have the inclusion $\hat{K} \subset (P)_\varepsilon$, which together with (6.3) yields the inequality

$$\operatorname{dist}(\hat{K}, P) < \varepsilon. \tag{6.4}$$

Clearly,

$$P \subset (\operatorname{cone}\{P\}) \cap U_r(0_m) \subset \hat{K}.$$

Using (6.4), the above inclusions and the equality $\operatorname{cone}\{P\} = \operatorname{cone}\{u^1, u^2, \ldots, u^l\}$, write

$$\operatorname{dist}(\hat{K}, \operatorname{cone}\{u^1, u^2, \ldots, u^l\} \cap U_r(0_m)) < \varepsilon. \tag{6.5}$$

By turn, it follows from (6.5) and $R^m_+ \subset K$ that

$$\operatorname{dist}(\hat{K}, \operatorname{cone}\{e^1, e^2, \ldots, e^m, u^1, u^2, \ldots, u^l\} \cap U_r(0_m)) < \varepsilon.$$

This inequality coincides with the required result (6.2) if we eliminate all "redundant" vectors from the collection $u^1, u^2, \ldots, u^l$ (i.e., the ones belonging to the nonnegative orthant R^m_+) and denote the residual vectors by $u^1, u^2, \ldots, u^k$. ■

[3] *The convex hull* of a given set A is the smallest convex set containing A.

6.3 Second Completeness Theorem

6.3.1 Example

Under definite conditions, the cone of the unknown preference relation can be approximated "from the inside" with an arbitrary accuracy by a polyhedral cone that corresponds to a certain finite collection of information quanta about this relation, see Sect. 6.2 for details. Note that the closeness of the cones of these relations [measured in terms of distance (6.1)] generally does not imply the closeness of the binary relations, ergo the closeness of the sets of nondominated vectors constructed using these relations. A simple example below confirms this fact.

Example 6.1 Choose $m = 2$, consider the two-dimensional set of feasible vectors (points) of the form

$$Y = \{(y_1, y_2) \in R^2 | y_1, y_2 \geqq 0, y_1 + y_2 = 1\},$$

and define the acute convex cone K by

$$K = (\text{cone}\{(1,0),(-1,1)\}) \backslash \{0_m\}.$$

Here the point $(0,1) \in Y$ dominates all other points from the segment (see highlighting in bold type in Fig. 6.1) in terms of the cone relation with the cone K. In particular, we have the relationship $(0,1) \succ (1,0)$, since $(0,1) - (1,0) = (-1,1) \in K$.

Now, slightly change K by considering the cone

$$K_\varepsilon = (\text{cone}\{(1,0),(-1,1+\varepsilon)\}) \backslash \{0_m\},$$

where $\varepsilon \in (0,1)$ (see Fig. 6.1). Using a sufficiently small positive number ε, the cone K_ε can be made arbitrarily close to the cone K in the sense of distance (6.1).

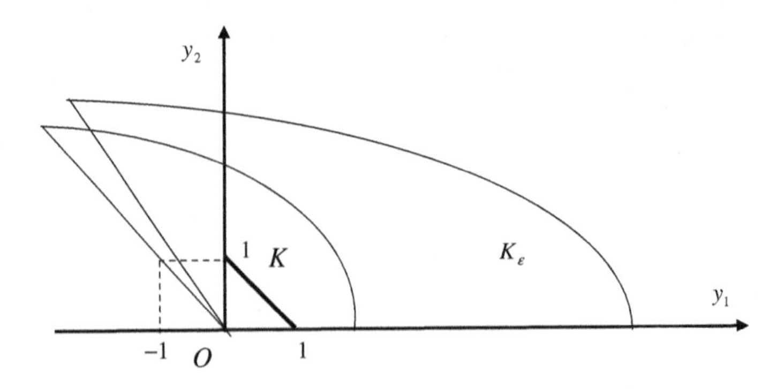

Fig. 6.1 Cones K and K_ε

On the other hand, the cone relation with the cone K_ε does not appear close to the cone relation with the cone K regardless of the choice of ε. For the latter relation, the set of nondominated points consists of the point $(0,1)$ only, whereas for the former relation the set of nondominated points is the whole segment between the points $(0,1)$ and $(1,0)$ for any $\varepsilon \in (0,1)$.

6.3.2 Second Completeness Theorem

As a matter of fact, Example 6.1 shows the following. If the cone K is not an open set, then a "small" change of this cone may modify appreciably the corresponding set of nondominated points. However, within the class of preference relations with open cones, the set of nondominated points in terms of an arbitrarily relation satisfying the conditions of Theorem 6.1 can be obtained as the limit of the sequence of the sets of nondominated points in terms of some cone relations constructed using a collection of information quanta. To be more precise, the following result takes place.

Theorem 6.2. *Let K be an open acute convex cone without the origin, $K \supset R^m_+$ and $K \neq R^m_+$. Assume that the set Y is K-bounded. Then there exists a vector sequence*

$$\{u^s\}_{s=1}^{\infty}, \quad u^s \in N^m \cap K, \quad s = 1,2,\ldots,$$

with rational components such that

$$\text{Ndom}_{\succ_s} Y \to \text{Ndom}\, Y \text{ as } s \to \infty, \tag{6.6}$$

where $\succ_s$ is the cone relation induced by the acute convex $\text{cone}\{e^1, e^2, \ldots, e^m, u^1, u^2, \ldots, u^s\}$ *without the origin, $s = 1, 2, \ldots$.*

Remark 6.1 In formula (6.6), we mean the so-called *pointwise convergence* of the sets of nondominated vectors defined in the following way. A point (vector) $y^* \in Y$ belongs to the limit set, i.e., $y^* \in \text{Ndom}\, Y$, if and only if there exists a natural number s_0 such that the inclusion $y^* \in \text{Ndom}_{\succ_s} Y$ holds for all natural numbers $s > s_0$.

□ Choose $\varepsilon = 1/n$. Using the proof of Theorem 6.1, for $n = 1$ we establish the existence of a vector collection $u^1, u^2, \ldots, u^k$ such that

$$d_r(K, \text{cone}\{e^1, e^2, \ldots, e^m, u^1, u^2, \ldots, u^k\}) < 1.$$

For $n = 2$, by analogy there exists another vector collection $u^{k+1}, \ldots, u^{k+p}$ such that

$$d_r(K, \text{cone}\{e^1, e^2, \ldots, e^m, u^{k+1}, u^{k+2}, \ldots, u^{k+p}\}) < 1/2.$$

The "extension" of the cone$\{u^{k+1}, \ldots, u^{k+p}\}$ by adding the obtained vectors $u^1, \ldots, u^k \in K$ does not increase the distance between K and the "extended" cone$\{u^1, \ldots, u^k, u^{k+1}, \ldots, u^{k+p}\}$. Hence,

$$d_r(K,\ \text{cone}\{e^1, e^2, \ldots, e^m, u^1, u^2, \ldots, u^k, u^{k+1}, u^{k+2}, \ldots, u^{k+p}\}) < 1/2$$

The same line of reasoning proves the existence of a vector sequence $\{u^s\}_{s=1}^{\infty}$ with the following property. For each natural number n, there is a number s_n such that

$$d_r(K, \text{cone}\{e^1, e^2, \ldots, e^m, u^1, u^2, \ldots, u^s\}) < 1/n,\ s = s_n, s_n + 1, \ldots, \tag{6.7}$$

Introduce the convex closed cones

$$C_s = \text{cone}\{e^1, e^2, \ldots, e^m, u^1, u^2, \ldots, u^s\},\ s = 1, 2, \ldots,$$

Clearly, $C_s \subset C_{s+1} \subset K \cup \{0_m\}$, $s = 1, 2, \ldots$ Moreover, by inequality (6.7) for any n there exists a number s_n such that

$$d_r(K, C_s) < \frac{1}{n}, \quad s = s_n, s_n + 1, \ldots.$$

This inequality immediately implies that, for any point $z \in \text{int}\,\hat{K}$ where $\hat{K} = K \cap U_r(0_m)$, there exists a number s_0 under which the inclusion $z \in C_s$ holds for all $s = s_0, s_0 + 1, \ldots$.

Now, we demonstrate convergence (6.6) for the nondominated sets. If $y^* \in Y$ and $y^* \notin \text{Ndom}Y$, by the definition of a set of nondominated points there exists a point $y \in Y$ such that $y - y^* \in K$. Since the set Y is K-bounded, it follows that $y - y^* \in \hat{K}$. Next, K is an open cone and we may assume that $z = y - y^* \in \text{int}\,\hat{K}$ (otherwise, choose, e.g., $0.5\,(y - y^*) \in \text{int}\,\hat{K}$ as such an inner point z). According to the aforesaid, then there exists a number s_0 such that the inclusion $z = y - y^* \in C_s$ holds for all numbers $s = s_0, s_0 + 1, \ldots$ This implies $y^* \notin \text{Ndom}_{\succ_s} Y$ for all s mentioned. And so, each point of the set Y that does not belong to the set Ndom$Y$ is not the limit point of the sequence of the sets $\text{Ndom}_{\succ_s} Y$, $s = 1, 2, \ldots$.

On the other hand, each point from Ndom$Y$ surely belongs to the above limit of the sequence, since the inclusions $C_s \subset C_{s+1} \subset K \cup \{0_m\}$ yield $\text{Ndom}_{\succ_s} Y \supset \text{Ndom}\,Y$ for all $s = 1, 2, \ldots$.

This concludes the proof of formula (6.6) and Theorem 6.2. ■

6.3.3 Case of Finite Set of Feasible Vectors

If the set of feasible vectors Y is finite, it suffices to use a definite finite collection of information quanta to find precisely the set of nondominated vectors in terms of the cone relation with an open cone K. The next theorem states this fact formally, being of great importance for the approach developed in the book. Really, Theorem 6.3 gives a theoretical justification for the axiomatic approach in the design of the set of nondominated vectors. According to this theorem, it is possible to find the set of nondominated vectors for the multicriteria choice problems of a definite class using only information quanta about the DM's preference relation.

Theorem 6.3 *Under the hypotheses of Theorem 6.2, let the set of feasible vectors Y be finite.* [4] *Then there exists a collection of p vectors $\{u^i\}_{i=1}^p \subset N^m \cap K$ with rational components such that*

$$\mathrm{Ndom}Y = \mathrm{Ndom}_{\succ_p} Y,$$

where $\succ_p$ is the cone relation induced by the acute convex cone$\{e^1, e^2, \ldots, e^m, u^1, u^2, \ldots, u^p\}$ *without the origin.*

Let $Y = \{y^1, y^2, \ldots, y^N\}$. For each $y^i \in Y$, introduce the finite set

$$Z_i = \{z \in Y \mid \text{there exists}\, y \in Y \text{ such that}\, z = y - y^i \in K\}, \quad i = 1, 2, \ldots, N.$$

Since K is an open set, there exists a number s_i such that $Z_i \subset C_s$ for all $s = s_i, s_i + 1, \ldots$, where C_s denote the cones defined in the proof of Theorem 6.2. Choose $p = \max\{s_1, s_2, \ldots, s_N\}$. Due to the inclusions $C_s \subset C_{s+1}, s = 1, 2, \ldots$, we obtain

$$Z_i \subset C_p \text{ for all } i = 1, 2, \ldots, N.$$

Take two arbitrary vectors $y^i, y^j \in Y$, $y^i \neq y^j$. If $y^j - y^i \in K$, then $y^j - y^i \in Z_i \subset C_p$ and hence $y^j - y^i \in C_p$. Conversely, the inclusion $y^j - y^i \in C_p$ together with $C_p \subset K \cup \{0_m\}$ leads to the inclusion $y^j - y^i \in K$. Therefore, we have established the equivalence

$$y^j - y^i \in K \Leftrightarrow y^j - y^i \in C_p,$$

which proves the equality of the cone relations with the cones K and $C_p \backslash \{0_m\}$. ■

[4] The K-boundedness of the set of feasible vectors can be omitted, since any finite set is bounded, ergo K-bounded.

Chapter 7
Pareto Set Reduction Using Fuzzy Information

In a series of applications-relevant multicriteria choice problems, the available information about the DM's preference relation can be fuzzy in the sense that it is impossible to define explicitly the preference for one alternative rather than another, since there exist the pros and cons of it. Then the framework of fuzzy sets and relations gives a convenient mathematical tool to describe such preference relations.

The current chapter formulates the multicriteria choice problem with a fuzzy preference relation, as well as introduces the notion of a fuzzy information quantum. In addition, the consistency of a finite collection of fuzzy information quanta and their consideration for Pareto set reduction are analyzed.

7.1 Statement of Fuzzy Multicriteria Choice Problem

7.1.1 Basic Notions from Theory of Fuzzy Sets

Recall some major notions from theory of fuzzy sets. The details can be found, e.g., in the books [20, 21].

Let A be a certain non-empty set (the so-called *universal* set). *A fuzzy set* X in A is defined by a membership function $\lambda_X : A \to [0, 1]$. For each element $x \in A$, the number $\lambda_X(x) \in [0, 1]$ is interpreted as its *grade of membership* to the set X. Speaking about a given fuzzy set, quite often researchers refer to its membership function, as the latter uniquely defines the corresponding fuzzy set. If the membership function $\lambda_X(\cdot)$ takes values 0 and 1 only, it becomes the characteristic function of a common (crisp) set X. All elements x of the set A that satisfy $\lambda_X(x) > 0$ form the *support* of the set X, further denoted by $\operatorname{supp} X$. Two fuzzy sets are equal to each other if they have the identical membership functions.

For two fuzzy sets X and Y, the inclusion relation, the operations of union and intersection are defined (in terms of membership functions) in the following way:

© Springer International Publishing AG 2018

V.D. Noghin, *Reduction of the Pareto Set*, Studies in Systems, Decision and Control 126, https://doi.org/10.1007/978-3-319-67873-3_7

$$X \subset Y \Leftrightarrow \lambda_X(x) \leqq \lambda_Y(x) \quad \text{for all } x \in A,$$
$$\lambda_{X \cup Y}(x) = \max\{\lambda_X(x); \lambda_Y(x)\} \quad \text{for all } x \in A,$$
$$\lambda_{X \cap Y}(x) = \min\{\lambda_X(x); \lambda_Y(x)\} \quad \text{for all } x \in A.$$

Consider a fuzzy set defined by a membership function $\eta(\cdot)$ in some linear space L. It forms

- *a fuzzy cone* if the equality $\eta(x) = \eta(\alpha \cdot x)$ holds for all $\alpha > 0$ and all $x \in L$;
- *a fuzzy acute cone* if the support of this cone is acute, i.e., none of the nonzero elements is contained in the support together with the opposite element;
- *a fuzzy convex set* if the inequality $\eta(\theta x + (1 - \theta) y) \geqq \min\{\eta(x); \eta(y)\}$ takes place for all $x, y \in L$ and all $\theta \in [0, 1]$.

A fuzzy binary relation is defined on the set A using a membership function $\mu : A \times A \to [0, 1]$. The number $\mu(x, y) \in [0, 1]$ is interpreted as the degree of confidence in that an element x has this relation with an element y. All pairs $(x, y) \in A$ that satisfy $\mu(x, y) = 1$ form the so-called *crisp part* of this fuzzy relation.

A fuzzy relation with a membership function $\mu(\cdot, \cdot)$ is called

- *irreflexive* if $\mu(x, x) = 0$ for all $x \in A$;
- *transitive* if $\mu(x, z) \geqq \min\{\mu(x, y); \mu(y, z)\}$ for all $x, y, z \in A$;
- *asymmetrical* if $\mu(x, y) > 0 \Rightarrow \mu(y, x) = 0$ for all $x, y \in A$;
- *a fuzzy cone relation* in a linear space L if there exists a fuzzy cone $\eta : L \to [0, 1]$ such that $\mu(x, y) = \eta(x - y)$ for all $x, y \in L$;
- *invariant with respect to a linear positive transformation* if this relation is defined in a linear space L and the equalities $\mu(\alpha x, \alpha y) = \mu(x, y)$, $\mu(x + c, y + c) = \mu(x, y)$ hold for all $x, y \in L$, $\alpha > 0$, $c \in L$.

As in crisp case, it is easy to verify that any irreflexive and transitive fuzzy relation is asymmetrical.

7.1.2 *Fuzzy Multicriteria Choice Problem*

Denote by X a (crisp) set of feasible alternatives that contains at least two elements. There are m ($m \geqq 2$) numerical functions $f_1, f_2, \ldots, f_m$ defined on the set X. They form the vector criterion $f = (f_1, f_2, \ldots, f_m)$ that takes values in the m-dimensional vector space R^m.

Suppose that the DM is not always able to decide unambiguously which of the two given alternatives is preferable. Accordingly, we consider a DM's fuzzy preference relation defined on the set X with a membership function $\mu_X(\cdot, \cdot)$. For alternatives $x', x'' \in X$, the number $\mu(x', x'') \in [0, 1]$ represents the degree of DM's confidence in that the alternative x' is preferable to x''.

Now, let us specify all elements of the fuzzy multicriteria choice problem $\langle X, f, \mu_X \rangle$ in terms of alternatives:

1. the set of feasible alternatives X,
2. a numerical vector criterion f defined on the set X,
3. a fuzzy preference relation with a membership function $\mu_X(\cdot,\ \cdot)$ that is defined on the Cartesian product $X \times X$ and takes values within the interval $[0,\ 1]$. In practice, this relation is known only fragmentary.

As is well-known, choice consists in indicating a certain alternative (selectable alternative) among all feasible alternatives, but in some cases a whole subset of alternatives is selected from the set X. It may appear difficult to judge whether certain alternatives are "good" or "bad" for choice. On the one hand, such alternatives have a series of advantages, which is a good reason to reckon them among the appropriate ones. On the other, these alternatives suffer from some shortcomings, which throws doubt upon their choice. In other words, the "good"-"bad" classification of all feasible alternatives seems "rough" in such situations. Here a more flexible and convenient approach involves theory of fuzzy sets when for each feasible alternative one has to assign a certain number from the interval $[0,\ 1]$. This number can be treated as the relative degree (or share) of positive or desirable qualities of the corresponding alternative. Therefore, generally the solution of the fuzzy multicriteria choice problem is a fuzzy set $C(X)$, $C(X) \subset X$, with a membership function μ_X^C. This set has to be found at the end of decision making process.

The above multicriteria choice problem can be stated in terms of vectors. Denote by $C(Y)$ the fuzzy set of selectable vectors whose membership function is naturally associated with the membership function of the fuzzy set of selectable alternatives as follows:

$$\lambda_X^C(Y) = \begin{cases} \lambda_X^C(x), & \text{if } y = f(x) \text{ for some } x \in X, \\ 0, & \text{if } y \in R^m \backslash Y. \end{cases}$$

Suppose that there exists a bijection between the set of feasible alternatives and the set of corresponding vectors. Then the membership function of the fuzzy set of selectable alternatives is defined through the membership function of the fuzzy set of selectable vectors.

A function $\mu_X(\cdot,\cdot)$ induces a membership function $\mu_Y(\cdot,\cdot)$ of a fuzzy preference relation on a set Y in the following way:

$$\mu_Y(y', y'') = \mu_X(x', x'') \quad \Leftrightarrow \quad y' = f(x'),\ y'' = f(x'') \quad \text{for all} \quad x', x'' \in X.$$

By-turn, the membership function of a fuzzy preference relation defined on Y induces the membership function of the fuzzy preference relation on the set X.

Consequently, the fuzzy multicriteria choice problem in terms of vectors includes

1. the set of feasible vectors Y,
2. a fuzzy preference relation with a membership function $\mu_Y(\cdot,\ \cdot)$ defined on Y.

This problem is to find the fuzzy set of selectable vectors $C(Y)$ with the membership function μ_Y^C.

The two fuzzy multicriteria choice problems formulated in terms of alternatives and in terms of vectors are equivalent to each other: owing to the bijection and compatibility of their membership functions, all results obtained in terms of one problem can be reformulated in terms of the other.

The solution of both problems entails difficulties, since in practice the DM's fuzzy preference relation is often unknown. This aspect essentially complicates the design of the set of selectable vectors (alternatives).

7.1.3 Axioms of Fuzzy Reasonable Choice

Here we provide several axioms that are accepted throughout Chap. 7. They actually represent the extensions of the corresponding axioms (see Chaps. 1 and 2) to the case of a fuzzy preference relation.

Axiom F1 (exclusion axiom). *For each pair of alternatives $x', x'' \in X$ that satisfies $\mu_X(x', x'') = \mu^* \in [0,\ 1]$, we have the inequality $\lambda_X^C(x'') \leqq 1 - \mu^*$. In other words, for all $x', x'' \in X$ the inequality $\lambda_X^C(x'') \leqq 1 - \mu_X(x', x'')$ holds.*

Axiom F2 (transitivity axiom). *A fuzzy preference relation with a membership function $\mu_X(\cdot,\cdot)$ (hence, with a corresponding membership function $\mu_Y(\cdot,\cdot)$) is irreflexive and transitive. Moreover, there exists an irreflexive and transitive relation defined in the all criterion space R^m with a membership function $\mu(\cdot,\cdot)$ such that its restriction to Y coincides with the preference relation $\mu_Y(\cdot,\cdot)$.*

We say that criterion f_i *is compatible with a preference relation* $\mu(\cdot,\cdot)$ if for any vectors $y', y'' \in R^m$ the relationships

$$y' = (y'_1, \ldots y'_{i-1}, y'_i, y'_{i+1}, \ldots, y'_m),$$
$$y'' = (y'_1, \ldots y'_{i-1}, y''_i, y'_{i+1}, \ldots, y'_m),$$
$$y'_i > y''_i,$$

imply the equality $\mu(y', y'') = 1$.

Just like in the case of a crisp preference relation, the compatibility of a given criterion with a preference relation means that the DM is interested in the largest possible values of this criterion, other things being equal.

Axiom F3 (compatibility axiom). *Each of the criteria $f_1, f_2, \ldots, f_m$ is compatible with the preference relation $\mu(\cdot,\cdot)$.*

Axiom F4 (invariance axiom). *The fuzzy preference relation $\mu(\cdot,\cdot)$ is invariant with respect to a linear positive transformation.*

As easily seen, the above axioms are transformed into their counterparts from Chaps. 1 and 2 if the fuzzy preference relation becomes crisp.

Denote by

$$\lambda_X^P(x) = \begin{cases} 1, & \text{if } x \in P_f(X), \\ 0, & \text{otherwise} \end{cases}$$

the membership function of the set of Pareto optimal alternatives (i.e., the characteristic function of this set). The membership function $\lambda_Y^P(y)$ of the set of Pareto optimal vectors is introduced by analogy.

7.1.4 Fuzzy Pareto Principle

S. Orlovsky [51] adopted the notion of the *fuzzy set of nondominated alternatives* with the membership function λ_X^N defined by the formula

$$\lambda_X^N(x) = 1 - \sup_{z \in X} \mu_X(z, x) \quad \text{for all } x \in X.$$

Let $\operatorname{Ndom} X$ designate the fuzzy set of nondominated alternatives. If the preference relation is crisp (i.e., the membership function $\mu_X(\cdot,\cdot)$ takes values 0 or 1 only), the fuzzy set of nondominated alternatives coincides with the standard (crisp) set of nondominated alternatives, see Chap. 1 for details.

Lemma 7.1 (in terms of alternatives). *The acceptance of Axiom* F1 *guarantees the inclusion*

$$C(X) \subset \operatorname{Ndom} X \tag{7.1}$$

for any fuzzy set of selectable alternatives $C(X)$.

□ Choose an arbitrary alternative $x \in X$. Owing to Axiom F1, for any $z \in X$ we have $\lambda_X^C(x) \leqq 1 - \mu_X(z, x)$. Passing to the greatest lower bound over $z \in X$ in the right-hand side of this inequality yields

$$\lambda_X^C(x) \leqq \inf_{z \in X}(1 - \mu_X(z, x)) = 1 - \sup_{z \in X} \mu_X(z, x) = \lambda_X^N(x) \quad \text{for all } x \in X.$$

The inequality $\lambda_X^C(x) \leqq \lambda_X^N(x)$ holding for all $x \in X$ proves inclusion (7.1). ■

According to the (crisp) Pareto axiom, the inequality $f(x'') \geq f(x')$ implies the relationship $x'' \succ_X x'$. For the fuzzy preference relation the corresponding version of *the Pareto axiom* is stated as follows: the inequality $f(x'') \geq f(x')$ implies the equality $\mu_X(x'', x') = 1$. This axiom can be formulated in terms of vectors by analogy.

Theorem 7.1 (fuzzy Pareto principle). *Under Axiom* F1 *and the Pareto axiom, the inclusion*

$$C(X) \subset P_f(X) \tag{7.2}$$

holds for any fuzzy set of selectable alternatives $C(X)$. *Equivalently, the relationship* $\lambda_X^C(x) \leqq \lambda_X^P(x)$ *holds for all* $x \in X$.

□ Owing to Lemma 7.1, we have the inclusion $C(X) \subset NdomX$, which is equivalent to the relationship $\lambda_X^C(x) \leqq \lambda_X^N(x)$ for all $x \in X$. It remains to check that the fuzzy set of nondominated alternatives Ndom X is a subset of the Pareto set $P_f(X)$. Conjecture the opposite, i.e., there exists $x' \in X$ such that $\lambda_X^N(x') > \lambda_X^P(x')$. Since the characteristic function of the Pareto set takes only two values (0 and 1), we arrive at the equality $\lambda_X^P(x') = 0$. The latter means that the alternative x' is not Pareto optimal and hence there exists $x'' \in X$ such that $f(x'') \geq f(x')$. According to the Pareto axiom, the last inequality implies $\mu(x'', x') = 1$. In this case, by Axiom F1 we get $\lambda_X^N(x') = 0$, and this result together with $\lambda_X^P(x') = 0$ contradicts the initial hypothesis $\lambda_X^N(x') > \lambda_X^P(x')$. ■

Inclusion (7.2) expresses the general principle of fuzzy choice: **in a rather wide class of fuzzy multicriteria choice problems** (the ones satisfying Axiom F1 and the Pareto axiom), **any choice (including fuzzy choice) must be made within the Pareto set.**

In the special case of a crisp preference relation, Theorem 7.1 coincides with the (crisp) Edgeworth–Pareto principle, see Chap. 1.

We accept Axioms F1–F3 for the remainder of this chapter. Similarly to the case of a crisp preference relation,

Axioms F2 *and* F3 *guarantee the Pareto axiom.*

□ Really, choose two arbitrary vectors $y', y'' \in R^m$ such that $y'' = f(x'') \geq f(x') = y'$. This vector inequality contains strict inequality for at least one component. Suppose that there are l such strict inequalities $(0 < l \leqq m)$ and all of them correspond to the first l indexes of the criteria (the last assumption does not restrict the generality of subsequent considerations). Then we have the chain of inequalities

$$\begin{aligned} y'' \geq y^1 &= (y_1', y_2'', \ldots, y_m'') \geq y^2 = (y_1', y_2', y_3'', \ldots, y_m'') \geq \ldots \geq y^l \\ &= (y_1', y_2', \ldots, y_l', y_{l+1}'', \ldots, y_m'') = y'. \end{aligned}$$

According to Axiom F3, it leads to the chain of equalities

$$\mu(y'', y^1) = \mu(y^1, y^2) = \ldots = \mu(y^{l-1}, y^l) = \mu(y^{l-1}, y') = 1,$$

which yields the desired result $\mu(y'', y') = 1$ due to the transitivity of the fuzzy preference relation. ■

7.2 Fuzzy Information About Preference Relation and Its Consistency

7.2.1 Definition and Some Properties of Information Quantum on Fuzzy Preference Relation

Definition 7.1 Consider two groups of criteria indexes, A and B, such that $A, B \subset I$, $A \neq \emptyset$, $B \neq \emptyset$, and $A \cap B = \emptyset$. We say that there is a *given quantum of fuzzy information with the groups of criteria* A *and* B,[1] *two collections of positive parameters* w_i *for all* $i \in A$ *and* w_j *for all* $j \in B$, *and a degree of confidence* $\mu^* \in (0, 1]$ if the equality $\mu(y', y'') = \mu^*$ holds for all vectors $y', y'' \in R^m$ satisfying

$$\begin{array}{ll} y'_i - y''_i = w_i & \text{for all } i \in A, \\ y''_j - y'_j = w_j & \text{for all } j \in B, \\ y'_s = y''_s & \text{for all } s \in I \backslash (A \cup B). \end{array}$$

As before, the number

$$\theta_{ij} = \frac{w_j}{w_i + w_j} \in (0, 1)$$

is called the *degree (coefficient) of compromise*.

If there exists a fuzzy information quantum as described by Definition 7.1, we say that the group of criteria A *is more important* than the group of criteria B with the corresponding parameters and degree of compromise.

The next proposition reveals the mathematical sense of the fuzzy choice axioms introduced earlier.

Lemma 7.2 *The two statements below are equivalent:*

1. *The fuzzy relation* $\mu(\cdot, \cdot)$ *satisfies Axioms* F2–F4;
2. *The fuzzy relation* $\mu(\cdot, \cdot)$ *is a cone relation with a fuzzy acute convex cone that contains the nonnegative orthant* R^m_+ *and does not contain the origin with the unit grade of membership.*

□ I. Establish that the invariance of the preference relation with the membership function $\mu(\cdot, \cdot)$ (which is postulated by Axiom F4) is equivalent to the conicity of this relation. Let μ be invariant. Introduce a fuzzy set η by

$$\eta(x) = \mu(x, 0_m) \quad \text{for all } x \in R^m.$$

[1] Recall that, whenever no confusion occurs, we refer to a group of criteria by specifying the group of their indexes.

Owing to invariance, we have

$$\eta(\alpha x) = \mu(\alpha x, \alpha 0_m) = \mu(x, 0_m) = \eta(x) \quad \text{for all } x \in R^m, \alpha > 0.$$

Hence, the fuzzy set η is a cone. In addition,

$$\mu(x, y) = \mu(x - y, 0_m) = \eta(x - y) \quad \text{for all } x, y.$$

This means that the invariant relation μ is a cone relation.

Conversely, let μ represent a fuzzy cone relation with a cone η. Its invariance follows from the equalities

$$\mu(x, y) = \eta(x - y) = \eta((x + a) - (y + a)) = \mu(x + a, y + a) \quad \text{for all } a \in R^m,$$
$$\mu(x, y) = \eta(x - y) = \eta(\alpha(x - y)) = \eta(\alpha x - \alpha y)) = \mu(\alpha x, \alpha y) \quad \text{for all } \alpha > 0.$$

II. We will prove the implication (1) $\Rightarrow$ (2). Assume that the fuzzy relation μ satisfies Axioms F2–F4. By the aforesaid, the relation μ is a cone relation; denote its cone by η. The support of the cone η does not contain the origin, since otherwise the relation would not be irreflexive. To demonstrate that the cone η is acute, conjecture the opposite, i.e., there exists a nonzero vector y such that $\mu(y, 0_m) = \eta(y) > 0$ and $\mu(0_m, y) = \eta(-y) > 0$. Based on the transitivity of the relation μ , we accordingly have $\mu(y, y) > 0$, which contradicts the irreflexivity of μ.

To show that the cone η is convex, take $x = \alpha x', y = 0_m, z = (1 - \alpha)(-x'')$, $\alpha \in (0, 1)$ in the definition of a transitive fuzzy relation. In this case,

$$\mu(\alpha x', (1 - \alpha)(-x'')) \geqq \min\{\mu(\alpha x', 0_m), \mu(0_m, (1 - \alpha)(-x''))\}.$$

Hence, using the invariance property, we get

$$\mu(\alpha x', (1 - \alpha)(-x'')) \geqq \min\{\mu(x', 0_m), \mu(0_m, -x'')\},$$

or

$$\eta(\alpha x' + (1 - \alpha)x'') \geqq \min\{\eta(x'), \eta(x'')\}.$$

This inequality means the convexity of the fuzzy cone η.

Let e^i be unit vector i in space R^m. Then Axiom F3 dictates that $\mu(e^i, 0_m) = \eta(e^i) = 1$. And so, all unit vectors of this space belong to the cone η with the unit grade of membership. Due to the convexity of η, the whole nonnegative orthant of this space has the same grade of membership.

Now, we prove the implication (2) $\Rightarrow$ (1). First of all, note that the fuzzy relation μ is invariant, as established earlier. Based on the convexity of the cone η, for any x, y, z we have

$$\eta\left(\frac{x-y}{2}+\frac{y-z}{2}\right) \geqq \min\{\eta(x-y), \eta(y-z)\}.$$

This gives the inequality

$$\mu(x,z) \geqq \min\{\mu(x,y), \mu(y,z)\} \quad \text{for all } x, y, z,$$

which establishes the transitivity of the fuzzy relation μ. The latter is also irreflexive, since the support of the cone η does not contain the origin. Therefore, the fuzzy relation μ satisfies Axioms F2 and F4. This relation obeys Axiom F3 as well: for any two vectors y' and y'' from the definition of a compatible criterion together with the invariance property we get

$$\mu(y', y'') = \mu(e^i, 0_m) = \eta(e^i) = 1.$$ ■

Lemma 7.3 *Under Axiom* F4, *the specification of a fuzzy information quantum with the groups of criteria A and B, given positive parameters w_i and w_j for all $i \in A$, $j \in B$, and the degree of confidence $\mu^* \in (0,\ 1]$ is equivalent to the equality $\mu(\tilde{y}, 0_m) = \mu^*$, where the vector $\tilde{y} \in R^m$ has the components $\tilde{y}_i = w_i$, $\tilde{y}_j = -w_j$, and $\tilde{y}_s = 0$ for all $i \in A, j \in B$, $s \in I \backslash (A \cup B)$.*

□ Necessity is obvious. To verify sufficiency, choose two arbitrary vectors $y', y'' \in R^m$ from Definition 7.1. Then the required result is immediate from the following equalities based on the invariance of the fuzzy relation μ:

$$\mu^* = \mu(\tilde{y}, 0_m) = \mu(\tilde{y} + y'', y'') = \mu(y', y'').$$ ■

As above, denote by N^m the set of all m-dimensional vectors having at least one positive and at least one negative components. According to Lemma 7.3, each vector from this set may define a certain quantum of fuzzy information if it is preferable to the zero vector with some nonzero degree of confidence.

7.2.2 Consistent Collection of Fuzzy Information Quanta

Consider a given collection of the pairs of vectors u^i, v^i $(u^i - v^i \in N^m)$ together with a collection of numbers $\mu_i \in (0,\ 1]$ such that $\mu(u^i, v^i) = \mu_i$, $i = 1, 2, \ldots, k$. Let

$$\mu_{11}, \ldots, \mu_{1k_1}; \mu_{21}, \ldots \mu_{2k_2}; \ldots; \mu_{l1}, \ldots, \mu_{lk_l}$$

be a permutation of the numbers $\mu_1, \mu_2, \ldots, \mu_k$ that satisfies

$$1 \geqq \mu_{11} = \cdots = \mu_{1k_1} > \mu_{21} = \cdots = \mu_{2k_2} > \cdots > \mu_{l1} = \cdots = \mu_{lk_l} > 0,$$

where $k_1 + \cdots + k_l = k, 1 \leqq l \leqq k$. Thus, we obtain the following bijection: each pair u^i, v^i corresponds to a definite positive number μ_{rs} ($r \in \{1, 2, \ldots, l\}$, $s \in \{1, 2, \ldots, k_r\}$) such that $\mu_i = \mu_{rs}$. Conversely, each number μ_{rs} above corresponds to a certain pair of vectors from the collection u^i, v^i, $i = 1, 2, \ldots, k$.

Let e^i be a unit vector of space R^m, $i = 1, 2, \ldots, m$. Introduce the crisp cones K_h, $h \in \{1, 2, \ldots, l\}$, generated by the unit vectors $e^1, e^2, \ldots, e^m$ together with the vectors $u^i - v^i$, $i \in \{1, 2, \ldots, k\}$, associated with the numbers μ_i of the form $\mu_i = \mu_{rs}$ for some r and s, and also $\mu_i \geqq \mu_{h1}$. This definition of cones K_h directly leads to the inclusions $K_1 \subset K_2 \subset \cdots \subset K_l$.

Definition 7.2 *A collection of the pairs of vectors* u^i, v^i $(u^i - v^i \in N^m)$, $i = 1, 2, \ldots, k$, *together with a collection of numbers* $\mu_1, \mu_2, \ldots, \mu_k \in (0,\ 1]$ *define a consistent collection of fuzzy information quanta if there exists at least one fuzzy preference relation* $\mu(\cdot, \cdot)$ *that satisfies Axioms* F2–F4 *and* $\mu(u^i, v^i) = \mu_i \in (0, 1]$, $i = 1, 2, \ldots, k$ $(k \geqq 1)$.

The following theorem gives a consistency criterion for a collection of fuzzy information quanta.

Theorem 7.2 *A collection of the pairs of vectors* u^i, v^i $(u^i - v^i \in N^m)$, $i = 1, 2, \ldots, k$, *together with a collection of numbers* $\mu_1, \mu_2, \ldots, \mu_k \in (0,\ 1]$ *specify a consistent collection of fuzzy information quanta if and only if the system of linear equations*

$$\lambda_1 e^1 + \cdots + \lambda_m e^m + \xi_1 (u^1 - v^1) + \cdots + \xi_k (u^k - v^k) = 0_m \tag{7.3}$$

has no N-solution $\lambda_1, \ldots, \lambda_m, \xi_1, \ldots, \xi_k$ *and, in addition, each cone* K_h, $h \in \{1, \ldots, l-1\}$, *contains no vectors* $u^i - v^i$, $i \in \{1, 2, \ldots, k\}$, *that are associated with a number* μ_i *satisfying the inequality* $\mu_i < \mu_{h1}$.

□ Necessity. Assume that a collection of the pairs of vectors u^i, v^i $(u^i - v^i \in N^m)$, $i = 1, 2, \ldots, k$, together with a collection of numbers $\mu_1, \mu_2, \ldots, \mu_k \in (0,\ 1]$ specify a consistent collection of fuzzy information quanta. Due to Definition 7.2 and Lemma 7.2, there exists a fuzzy preference relation μ with a fuzzy convex cone η, where the unit vectors $e^1, \ldots, e^m$ and the vectors $u^i - v^i$, $i = 1, 2, \ldots, k$, belong to the support of the cone η, which is a crisp acute cone. Hence, the crisp polyhedral convex cone generated by the vectors $e^1, \ldots, e^m$, $u^i - v^i$, $i = 1, 2, \ldots, k$, is also acute, and the system of linear Eqs. (7.3) has no N-solution $\lambda_1, \ldots, \lambda_m, \xi_1, \ldots, \xi_k$, see Remark 4.1.

To prove the concluding part of necessity, we conjecture the opposite, i.e., for some $h \in \{1, \ldots, l-1\}$ the cone K_h contains a vector $u^i - v^i$ whose grade of membership $\mu_i = \mu_{rs}$ satisfies the inequalities $\mu_{rs} < \mu_{h1}$, $r > h$, for some r and s. The inclusion $u^i - v^i \in K_h$ can be rewritten as

$$u^i - v^i = \sum_q \lambda_q e^q + \sum_j \alpha_j (u^j - v^j),$$

where $\lambda_q > 0,\ \alpha_j > 0$ and, in the second term, summation runs over the set of all j satisfying $\mu(u^j, v^j) \geqq \mu_{h1}$. Using the convexity of the fuzzy cone η, we have

$$\mu_{rs} = \mu_i = \mu(u^i, v^i) = \eta(u^i - v^i) = \eta\left(\sum_q \lambda_q e^q + \sum_j \alpha_j (u^j - v^j)\right) \geqq \min_{q,j}\{\eta(e^q), \eta(u^j - v^j)\}$$
$$= \min_j\{\eta(u^j - v^j)\} = \min_j\{\mu(u^j, v^j)\} = \mu_{t1}$$

for some $t \in \{1, \ldots, h-1\}$. Then the resulting inequalities $\mu_{rs} \geqq \mu_{t1}$ and $r > t$ contradict the accepted hypothesis:

$$1 \geqq \mu_{11} = \ldots = \mu_{1k_1} > \mu_{21} = \ldots = \mu_{2k_2} > \ldots > \mu_{l1} = \ldots = \mu_{lk_l} > 0.$$

Sufficiency. System (7.3) has no N-solution by the hypotheses of this theorem. And so, the crisp polyhedral convex cone K_l generated by the unit vectors $e^1, \ldots, e^m$ together with the vectors $u^i - v^i,\ i = 1, 2, \ldots, k,$ is acute according to the above-mentioned Remark 4.1

Consider the fuzzy cone relation μ with the fuzzy cone η that has the support K_l (without the origin), i.e.,

$$\eta(z) = \begin{cases} 1, & \text{if } z \in R^m_+, \\ \mu_i, & \text{if } z = \alpha(u^i - v^i) \text{ for some } \alpha > 0 \text{ and } i \in \{1, \ldots, k\}, \\ \max\limits_{j \in \{1,2,\ldots,l\}} \{\mu_{j1} | z \in K_j\}, & \text{in the rest cases.} \end{cases} \tag{7.4}$$

Obviously, $\mu(u^i, v^i) = \eta(u^i - v^i) = \mu_i,\ i = 1, \ldots, k$. And therefore it remains to show that the relation μ satisfies Axioms F2–F4. According to Lemma 7.3, this is equivalent to the following: the cone η is convex and contains the nonnegative orthant R^m_+ with the unit grade of membership and the origin with the zero grade of membership. Recall that (a) the support of the cone η is the cone K_l without the origin and (b) the cone η contains the nonnegative orthant with the unit grade of membership by (7.4). Hence, the only thing to do is to verify the convexity of the fuzzy cone η.

To this end, choose arbitrary vectors $x, y \in K_l$ and an arbitrary number $\lambda \in [0, 1]$. Without loss of generality, suppose that $x \in K_h, x \notin K_{h-1}$ and $y \in K_t, y \notin K_{t-1}$, where $l \geqq h \geqq t \geqq 0$ (note that $K_0 = R^m_+, K_{-1} = \emptyset$).

To prove that the cone η is convex, verify the inequality $\eta(z) \geqq \mu_{h1}$, where $z = \lambda x + (1 - \lambda) y$. By the hypothesis of this theorem, each cone $K_j,\ j \in \{1, \ldots, l\}$, contains the unit vectors $e^1, \ldots, e^m$ and also the vectors $u^i - v^i$ satisfying the inequality $\mu_i \geqq \mu_{j1}$. The cone K_h particularly contains the vectors $e^1, \ldots, e^m$ and only such vectors $u^i - v^i$ that satisfy $\mu_i \geqq \mu_{h1}$.

Let us verify the inequality $\eta(z) \geqq \mu_{h1}$. Since $K_h \supset K_t$ and the cone K_h is convex, we have $z \in K_h$. If the point z obeys the inclusion $z \in R^m_+$, then equality (7.4) leads to $\eta(z) = 1 \geqq \mu_{h1}$. If $z = \alpha(u^i - v^i)$ holds for some $\alpha > 0$ and $i \in \{1, \ldots, k\}$, then $\eta(z) = \mu_i \geqq \mu_{h1}$ due to the inclusion $z \in K_h$ and equality (7.4). In the rest cases, we obtain $z \in K_j$ and $z \notin K_{j-1}$ for some $j \in \{1, \ldots, h\}$, which also yield $\eta(z) = \mu_{j1} \geqq \mu_{h1}$ using equality (7.4). This proves the convexity of the cone η. ■

7.3 Pareto Set Reduction Based on Fuzzy Information Quantum

7.3.1 *Basic Result*

As a matter of fact, the next theorem extends Theorem 3.5 to the case of a fuzzy preference relation. Further exposition is based on Axioms F1–F4.

Theorem 7.3 *Let we have a given quantum of fuzzy information with the groups of criteria A and B, positive parameters w_i, w_j for all $i \in A$, $j \in B$, and the degree of confidence $\mu^* \in (0, 1]$. Then for any set of selectable vectors with a membership function $\lambda^C_Y(\cdot)$ the inequalities*

$$\lambda^C_Y(y) \leqq \lambda^M_Y(y) \leqq \lambda^P_Y(y) \quad \textit{for all } y \in Y, \tag{7.5}$$

hold, where $\lambda^P_Y(\cdot)$ is the membership function of the Pareto set and $\lambda^M_Y(\cdot)$ is the membership function defined by

$$\lambda^M_Y(y) = 1 - \sup_{z \in Y} \varsigma(z, y) \quad \textit{for all } y \in Y, \tag{7.6}$$

$$\zeta(z, y) = \begin{cases} 1, & \textit{if } z - y \in R^m_+, \\ \mu^*, & \textit{if } \hat{z} - \hat{y} \in R^p_+, \quad z - y \notin R^m_+, \\ 0, & \textit{in the rest cases}, \end{cases} \quad \textit{for all } y, z \in Y. \tag{7.7}$$

Here $p = m - |B| + |A| \cdot |B|$, while the vectors $\hat{y}$ and $\hat{z}$ have the components y_i and z_i, respectively, for all $i \in I \backslash B$, and $w_j y_i + w_i y_j$ and $w_j z_i + w_i z_j$, respectively, for all $i \in A$ and $j \in B$.

□ Owing to Lemma 7.3, the fuzzy preference relation is a cone relation with a fuzzy acute convex cone (without the origin). Denote it by K. This cone contains the nonnegative orthant R^m_+ with the unit grade of membership. The presence of an information quantum as stated by Lemma 7.3 means that the corresponding vector $\tilde{y}$ belongs to the cone K, and its grade of membership is μ^*.

Consider a fuzzy set M in the criterion space whose support represents the collection of all N-combinations for a finite collection of the unit vectors $e^1, e^2, \ldots, e^m$ from space R^m and the above vector $\tilde{y}$. The vectors of the support that belong to the nonnegative orthant have the unit grade of membership (except the origin with the zero grade of membership), while the rest vector of the support have the grade of membership μ^*. The set M is a fuzzy acute convex cone (without the origin) that contains the nonnegative orthant with the unit grade of membership and, owing to the convexity of the cone K, is a subset of this cone.

According to the results of Chap. 3, the support of the cone M coincides with the set of all solutions to the system of linear inequalities $\langle a^i, y \rangle \geq 0$, $i = 1, 2, \ldots, p = m - |B| + |A| \cdot |B|$, where the vectors $a^1, a^2, \ldots, a^p$ are e^i for all $i \in I \backslash B$ and $w_j e^i + w_i e^j$ for all $i \in A,\ j \in B$.

Consider the fuzzy cone relation ς with the cone M. Clearly, it coincides with the one defined by (7.7). On the other hand, the fuzzy set with the membership function $\lambda_Y^M(y)$ in inequality (7.5) represents the fuzzy set of nondominated vectors in terms of the relation ς, see formula (7.6).

We have mentioned that the nonnegative orthant belongs to the cone M. By-turn, this cone is contained in the cone K. The Pareto set therefore contains the set of all nondominated vectors in terms of the cone relation ς that includes the set of nondominated vectors generated by the cone preference relation with the membership function $\mu(\cdot, \cdot)$. And the latter set contains the fuzzy set of selectable vectors owing to the fuzzy Pareto principle. ■

Analysis of Theorem 7.3 shows the following. To construct the fuzzy set with the membership function $\lambda_Y^M(y)$ (i.e., to reduce the Pareto set based on the fuzzy information quantum), one has to solve two (crisp) multicriteria problems (more specifically, to find the Pareto sets in two multicriteria problems). First, it is necessary to consider the multicriteria problem incorporating the initial vector function f and the set of feasible alternatives X. And then assign the unit grade of membership to all vectors in the resulting Pareto set and the zero grade of membership to the rest vectors. Second, on the same set X it is necessary to solve the multicriteria problem with the new ("recalculated") p-dimensional vector function with the components f_i for all $i \in I \backslash B$ and $w_j f_i + w_i f_j$ for all $i \in A,\ j \in B$. And then assign the grade of membership $1 - \mu^*$ to all vectors in the "old" Pareto set that do not appear in the "new" Pareto set. The described procedure allows to reduce the Pareto set containing the unknown set of selectable vectors. An illustrative example will be given below.

In the special case where the sets A and B are singletons, Theorem 7.3 can be reformulated as follows.

Corollary 7.1 *Let $i, j \in I$, $i \neq j$, and consider a given elementary quantum of fuzzy information with the pair of criteria f_i, f_j, parameters w_i, w_j and the degree of confidence $\mu^* \in (0, 1]$. Then for any set of selectable vectors with the membership function $\lambda_Y^C(\cdot)$ we have inequalities (7.5), where the membership function $\lambda_Y^M(y)$ and the relation ς are defined by (7.6) and (7.7), respectively, $p = m$ and the m-*

dimensional vectors $\hat{y}$ *and* $\hat{z}$ *have the components* $y_1, \ldots, y_{j-1}, w_j y_i + w_i y_j, y_{j+1}, \ldots, y_m$ *and* $z_1, \ldots, z_{j-1}, w_j z_i + w_i z_j, z_{j+1}, \ldots, z_m$, *respectively.*

7.3.2 Example

Example 7.1 Consider an illustrative example. Under the hypotheses of Corollary 7.1, let $m = 2$, $f = (f_1, f_2)$, $Y = \{y^1, y^2, y^3, y^4\} \subset R^2$, $y^1 = (0,\ 3)$, $y^2 = (1,\ 1)$, $y^3 = (2,\ 1)$, and $y^4 = (4,\ 0)$. In this case, the set of Pareto optimal vectors consists of three elements, $\lambda_Y^P(y^1) = \lambda_Y^P(y^3) = \lambda_Y^P(y^4) = 1$, $\lambda_Y^P(y^2) = 0$, since $y^3 \geq y^2$.

Assume that there is a fuzzy information quantum, which states that criterion f_1 is more important than criterion f_2 with the parameters $w_i = 0.3$, $w_j = 0.7$ and the degree of confidence 0.6. Then using Corollary 7.1 we obtain

$$\hat{y}^1 = (0, 0.9), \hat{y}^2 = (1, 1), \hat{y}^3 = (2, 1.7), \hat{y}^4 = (4, 2.8),$$
$$\varsigma(y^1, y^2) = \varsigma(y^1, y^3) = \varsigma(y^1, y^4) = \varsigma(y^2, y^3) = \varsigma(y^2, y^4) = \varsigma(y^3, y^4) = 0,$$
$$\varsigma(y^2, y^1) = \varsigma(y^3, y^1) = \varsigma(y^4, y^1) = \varsigma(y^4, y^2) = \varsigma(y^4, y^3) = 0.6, \varsigma(y^3, y^2) = 1,$$
$$\lambda_Y^M(y^1) = 1 - \max\{0.6, 0.6, 0.6\} = 0.4, \lambda_Y^M(y^2) = 1 - \max\{0, 1, 0.6\} = 0,$$
$$\lambda_Y^M(y^3) = 1 - \max\{0, 0, 0.6\} = 0.4, \lambda_Y^M(y^4) = 1 - \max\{0, 0, 0\} = 1.$$

The fuzzy set with the membership function $\lambda_Y^M(y^1) = 0.4$, $\lambda_Y^M(y^2) = 0$, $\lambda_Y^M(y^3) = 0.4$, $\lambda_Y^M(y^4) = 1$ yields an upper estimate for the unknown set of selectable vectors.

The same estimate can be constructed in a different way. At step 1, in the initial multicriteria problem assign the unit grade of membership to all Pareto optimal vectors and the zero grade of membership to the rest vectors (in the current example, to the vector y^2 only). At step 2, find the Pareto optimal vectors in the set of "recalculated" vectors $\{\hat{y}^1, \hat{y}^3, \hat{y}^4\}$. Such is the vector $\{\hat{y}^4\}$ only. Hence, the first and third vectors (y^1 and y^3) receive the grade of membership defined by $1 - 0.6 = 0.4$. This procedure yields the same upper estimate as before.

The following recommendations for choice can be suggested using the above example. If exactly one vector must be selected, in this case the DM should set its choice on the vector y^4. Analyze the same example under the condition that it is necessary to choose two vectors from the four ones. In this case, a possible approach is to consider the new problem where the feasible vectors are the pairs of vectors (instead of separate vectors). In this problem the preference relation is a new binary relation defined on these pairs of vectors. Thereby, we obtain a distinctly different (and much more difficult) problem, and its solution requires new information about the preference relation in comparison with the initial problem. Moreover, in this case the initial information is of little interest for subsequent choice.

Now, let us study the second approach to choose two vectors. If the individual properties of the alternatives combined in into pairs remain almost the same, then we may suggest the following recommendations within the framework of the initial problem. One should choose the pair containing the third and fourth vectors (y^3 and y^4), since the third vector is preferable to the first vector ($\hat{y}^3 \geq \hat{y}^1$) taking into account the fuzzy information quantum available.

Finally, if it is necessary to choose three vectors among the four ones, a natural recommendation consists in choosing the triplet that forms the Pareto set in the initial problem (i.e., the first, third and fourth vectors), since the (crisp) Edgeworth–Pareto principle holds under these conditions. Note that the existing information about the preference relation is not used at all; it becomes "redundant."

7.3.3 Case of a Fuzzy Set of Feasible Alternatives

The results established for a crisp set X can be extended to the case of a fuzzy set of feasible alternatives. Denote by $\lambda_X(\cdot)$ and $\lambda_Y(\cdot)$ the corresponding membership functions for alternatives and for vectors, respectively.

To carry out such extension, first we have to agree what is the solution of the fuzzy choice problem. Fuzzy choice is performed within the fuzzy set X, and so for each selectable alternative the grade of membership must not exceed its grade of membership in the fuzzy set X. We may therefore assume that *the fuzzy set of selectable alternatives* (further designated by $FC(X)$) with a membership function $\lambda_Y^F(\cdot)$ is by definition the intersection $X \cap C(X)$, where $C(X)$ indicates the set of selectable alternatives under the hypothesis that the set X is crisp. Thus, to solve the fuzzy choice problem with the fuzzy set X, it is necessary (1) to find the set of selectable alternatives under the hypothesis that the set X is crisp set and (2) to obtain the intersection of the resulting set with the given fuzzy set X.

The following theorems can be formulated using the aforesaid and the results established above.

Theorem 7.4 *In the case of the fuzzy set X and the fuzzy preference relation, for any fuzzy set of selectable alternatives* $FC(X)$ *we have the inclusion*

$$FC(X) \subset X \cap P_f(X).$$

Theorem 7.5 *Let the set X be fuzzy. If there is a given quantum of fuzzy information with the groups of criteria A and B, parameters* w_i, w_j *for all* $i \in A$, $j \in B$, *and the degree of confidence* $\mu^* \in (0, 1]$, *then for any set of selectable vectors with the membership function* $\lambda_Y^F(\cdot)$ *we have the inequalities*

$$\lambda_Y^F(y) \leqq \min\{\lambda_Y^M(y), \lambda_Y(y)\} \leqq \min\{\lambda_Y^P(y), \lambda_Y(y)\} \quad \text{for all } y \in Y,$$

where $\lambda_Y^P(\cdot)$ *is the membership function of the Pareto set,* $\lambda_Y^M(\cdot)$ *is the membership function defined by* (7.6), (7.7), *while the number p and the vectors* $\hat{y}$, $\hat{z}$ *are the same as in Theorem 7.4.*

7.4 Pareto Set Reduction Based on Collections of Fuzzy Information Quanta

7.4.1 *Pareto Set Reduction Using Two Fuzzy Information Quanta*

The next theorem deals with the consideration of two fuzzy information quanta stating that the criterion f_i is more important than the criteria f_j and f_k. Recall that $\mu(\cdot,\cdot)$ denotes the membership function of the preference relation from Axiom F2.

Theorem 7.6 *Let* $i, j, k \in I,\ i \neq j,\ i \neq k,\ j \neq k,$ *and* $\mu(y', 0_m) = \mu_1 \in (0,1]$, $\mu(y'', 0_m) = \mu_2 \in (0,1]$, *where* $\mu_1 \geqq \mu_2$ *and the vectors* y' *and* y'' *both have only two nonzero components,* $y'_i = w'_i,\ y'_j = -w'_j$ *and* $y''_i = w''_i,\ y''_k = -w''_k$, *respectively. Then for any set of selectable vectors with the membership function* $\lambda^C_Y(\cdot)$ *we have the inequalities*

$$\lambda^C_Y(y) \leqq \lambda^M_Y(y) \leqq \lambda^P_Y(y) \quad \textit{for all}\ y \in Y, \tag{7.8}$$

where $\lambda^P_Y(y)$ *is the membership function of the Pareto set and* $\lambda^M_Y(y)$ *is the membership function defined by*

$$\lambda^M_Y(y) = 1 - \sup_{z \in Y} \varsigma(z, y) \quad \textit{for all}\ y \in Y, \tag{7.9}$$

$$\varsigma(z, y) = \begin{cases} 1, & \textit{if}\ z - y \in R^m_+, \\ \mu_1, & \textit{if}\ \bar{z} - \bar{y} \in R^m_+,\ z - y \notin R^m_+, \\ \mu_2, & \textit{if}\ \hat{z} - \hat{y} \in R^{m+1}_+,\ \bar{z} - \bar{y} \notin R^m_+,\ z - y \notin R^m_+, \\ 0, & \textit{in the rest cases}, \end{cases} \quad \textit{for all}\ y, z \in Y, y \neq z, \tag{7.10}$$

and

$$\begin{aligned} \bar{y} &= \left(y_1, \ldots, y_{j-1}, w'_j y_i + w'_i y_j, y_{j+1}, \ldots, y_m\right), \\ \hat{y} &= (y_1, \ldots, y_{j-1}, w'_j y_i + w'_i y_j, y_{j+1}, \ldots, y_{k-1}, w''_k y_i \\ &\quad + w''_i y_k, y_{k+1}, \ldots, y_m, w'_j w''_k y_i + w'_i w''_k y_j + w''_i w'_j y_k), \\ \bar{z} &= \left(z_1, \ldots, z_{j-1}, w'_j z_i + w'_i z_j, z_{j+1}, \ldots, z_m\right), \\ \hat{z} &= (z_1, \ldots, z_{j-1}, w'_j z_i + w'_i z_j, z_{j+1}, \ldots, z_{k-1}, w''_k z_i \\ &\quad + w''_i z_k, z_{k+1}, \ldots, z_m, w'_j w''_k z_i + w'_i w''_k z_j + w''_i w'_j z_k). \end{aligned}$$

□ Owing to Lemma 7.2, each fuzzy relation $\mu(\cdot,\cdot)$ defined in the criterion space R^m that satisfies Axioms F2–F4 is a cone relation with the fuzzy acute convex cone

K without the origin that contains the nonnegative orthant with the unit grade of membership.

By the hypothesis of this theorem, we have two quanta and hence $k = 2$. Consider the two crisp cones K_1 and K_2 in the notation of Sect. 7.2.2. They satisfy the inclusion $K_1 \subset K_2$.

The existence of these two quanta means that the vector y' belongs to the cone K_1 with the grade of membership $\mu_1 = \mu(y', 0_m)$, while the vector y'' to the cone K_2 with the grade of membership $\mu_2 = \mu(y'', 0_m)$.

First, using Theorem 7.2 we verify that this collection of information quanta is consistent. The system of linear Eq. (7.3) acquires the form $\lambda_1 e^1 + \ldots + \lambda_m e^m + \xi_1 y' + \xi_k y'' = 0_n$. Equation i of this system gives $\lambda_i = \xi_1 = \xi_2 = 0$. Then it follows from equations j and k that $\lambda_j = \lambda_k = 0$. And the rest equations of this system dictate that all other components of the vector $\lambda = (\lambda_1, \lambda_2, \ldots, \lambda_m)$ are zero. Hence, the system of linear equations has no N-solution $\lambda_1, \ldots, \lambda_m, \xi_1, \xi_2$.

If $\mu_1 = \mu_2$, then obviously the collection of information quanta is consistent.

Suppose that $\mu_1 > \mu_2$, and verify the second condition of Theorem 7.2. By conjecturing the opposite, i.e.,$y'' \in K_1$, we may represent the vector y'' as N-combination of the vectors $e^1, \ldots, e^m, y'$. Equation k of the vector equality $\lambda_1 e^1 + \ldots + \lambda_m e^m + \xi_1 y' = y''$ leads to the inequality $\lambda_k < 0$, which contradicts the nonnegative property of the coefficients $\lambda_1, \ldots, \lambda_m$. And so, by Theorem 7.2 the collection of information quanta under consideration is consistent.

Introduce two fuzzy cones with the supports K_1 and K_2. Denote by M_1 the fuzzy cone with the support K_1 and the unit grade of membership for all its elements that belong to the nonnegative orthant and the grade of membership μ_1 for the other elements of this cone. Similarly, designate by M_2 the fuzzy cone with the support K_2 and the unit grade of membership for all its elements that belong to the nonnegative orthant and the grade of membership μ_2 for the other elements of this cone.

Let M be the fuzzy cone (without the origin) whose support is generated by the vectors $e^1, \ldots, e^m, y', y''$, so that M is the union of the fuzzy cones M_1 and M_2. Since the fuzzy information specified by the conditions of Theorem 7.6 is consistent, the fuzzy cone M is acute and convex. Thereby, the vectors of $\operatorname{supp} M$ $(= K_2)$ belonging to the nonnegative orthant have the unit grade of membership (except the origin with the zero grade of membership); the vectors belonging to the cone K_1 and not belonging to the nonnegative orthant have the grade of membership μ_1; the vectors belonging to the cone K_2 and not belonging to the cone K_1 have the grade of membership μ_2.

It follows from the proof of Theorem 4.5 that the support of the cone M coincides with the set of all nonzero solutions to the system of linear homogeneous inequalities

$$
\begin{gathered}
y_s \geqq 0 \quad \text{for all } s \in I \backslash \{j, k\}, \\
w'_j y_i + w'_i y_j \geqq 0, \\
w''_k y_i + w''_i y_k \geqq 0, \\
w'_j w''_k y_i + w'_i w''_k y_j + w''_i w'_j y_k \geqq 0,
\end{gathered}
$$

while supp M_1 coincides with the set of all solutions to the system of linear homogeneous inequalities

$$y_s \geqq 0 \quad \text{for all } s \in I \backslash \{j\},$$
$$w'_j y_i + w'_i y_j \geqq 0.$$

Consider the fuzzy cone relation with the cone M. Denote by ς its membership function. Analysis shows that this function is exactly the one defined by (7.10). The definition of the fuzzy set of nondominated vectors (7.9) and formula (7.10) imply that the fuzzy set with the membership function $\lambda_Y^M(y)$ is the fuzzy set of nondominated vectors in terms of the relation ς.

The nonnegative orthant is contained in M with the unit grade of membership. By-turn, the cone M is contained in the cone K of the fuzzy relation $\mu(\cdot,\cdot)$. The Pareto set hence includes the set of all nondominated vectors in terms of the cone relation ς, which includes the set of all nondominated vectors in terms of the cone relation M. Due to Lemma 7.1, the latter set contains any fuzzy set of selectable vectors. ■

According to Theorem 7.6, one has to solve three (crisp) multicriteria problems in order to construct the fuzzy set with the membership function $\lambda_Y^M(\cdot)$. First, it is necessary to find the Pareto set for the multicriteria problem incorporating the initial vector function f and the set of feasible alternatives X. And then assign the unit grade of membership to all vectors in the resulting Pareto set and the zero grade of membership to the rest vectors. Second, on the same set X it is necessary to solve the multicriteria problem with the new vector function with the components f_s for all $s \in I \backslash \{j\}$ and $w'_j f_i + w'_i f_j$. And then assign the grade of membership $1 - \mu_1$ to all vectors in the "old" Pareto set that do not appear in the "new" Pareto set. Third, on the same set X it is necessary to solve the multicriteria problem with the new vector function with the components f_s for all $s \in I \backslash \{j, k\}$, $w'_j f_i + w'_i f_j$, $w''_k f_i + w''_i f_k$, and the additional component $f_{m+1} = w'_j w''_k f_i + w'_i w''_k f_j + w'_j w''_i f_k$. And then assign the grade of membership $1 - \mu_2$ to all vectors in the "old" Pareto set that do not appear in the "new" Pareto sets for the second and third multicriteria problems. This procedure allows constructing a fuzzy set of vectors that is the reduced original Pareto set based on the two available quanta of fuzzy information.

7.4.2 Example

Example 7.2 Under the hypotheses of Theorem 7.6, let $m = 3$, $f = (f_1, f_2, f_3)$, $Y = \{y^1, y^2, y^3, y^4, y^5, y^6\} \subset R^3$, $y^1 = (4, 3, 5)$, $y^2 = (0, 3, 2)$, $y^3 = (1, 2, 3)$, $y^4 = (4, 3, 0)$, $y^5 = (5, 2, 7)$, and $y^6 = (2, 5, 5)$. In this case, the set of Pareto optimal vectors consists of three elements, since $\lambda_Y^P(y^1) = \lambda_Y^P(y^5) = \lambda_Y^P(y^6) = 1$ and $\lambda_Y^P(y^2) = \lambda_Y^P(y^3) = \lambda_Y^P(y^4) = 0$.

Assume that there is a fuzzy information quantum, which states that criterion f_1 is more important than criterion f_2 with the parameters $w_1' = 0.4$, $w_2' = 0.6$ and the degree of confidence 0.6, and that criterion f_1 is more important than criterion f_3 with the parameters $w_1'' = 0.5$, $w_3'' = 0.5$ and the degree of confidence 0.4.

Using Theorem 7.6 we obtain

$$\begin{array}{lll} \bar{y}^1 = (4, 3.6, 5), & \bar{y}^2 = (0, 1.2, 2), & \bar{y}^3 = (1, 1.4, 3), \\ \bar{y}^4 = (4, 3.6, 0), & \bar{y}^5 = (5, 3.8, 7), & \bar{y}^6 = (2, 3.2, 5). \end{array}$$

Here the set of Pareto optimal vectors consists of two elements, namely, the first and fifth vectors. Therefore,

$$\lambda_Y^M(y^1) = \lambda_Y^M(y^5) = 1, \quad \lambda_Y^M(y^6) = 0.4, \quad \lambda_Y^M(y^2) = \lambda_Y^M(y^3) = \lambda_Y^M(y^4) = 0.$$

According to Theorem 7.6, next we calculate the additional criterion $f_4 = 0.3f_1 + 0.2f_2 + 0.3f_3$. Hence,

$$\begin{array}{lll} \hat{y}^1 = (4, 3.6, 4.5, 3.3), & \hat{y}^2 = (0, 1.2, 1, 1.2), & \hat{y}^3 = (1, 1.4, 2, 1.6), \\ \hat{y}^4 = (4, 3.6, 2, 1.8), & \hat{y}^5 = (5, 3.8, 6, 4), & \hat{y}^6 = (2, 3.2, 3.5, 3.1). \end{array}$$

The fifth vector is Pareto optimal in this collection. In the end, we have

$$\lambda_Y^M(y^5) = 1, \lambda_Y^M(y^1) = 0.6, \lambda_Y^M(y^6) = 0.4, \quad \lambda_Y^M(y^2) = \lambda_Y^M(y^3) = \lambda_Y^M(y^4) = 0.$$

The resulting fuzzy set forms the desired reduction of the initial Pareto set. The best candidate for choice is the fifth vector, followed by the first and sixth vectors.

The next theorem shows how to reduce the Pareto using the two fuzzy information quanta stating that each of criteria f_i, f_j separately is more important than the criterion f_k.

Theorem 7.7 *Let* $i, j, k \in I$, $i \neq j$, $i \neq k$, $j \neq k$, *and* $\mu(y', 0_m) = \mu_1 \in (0, 1]$, $\mu(y'', 0_m) = \mu_2 \in (0, 1]$, *where* $\mu_1 \geqq \mu_2$ *and the vectors* y' *and* y'' *both have only two nonzero components,* $y_i' = w_i'$, $y_k' = -w_k'$ *and* $y_j'' = w_j''$, $y_k'' = -w_k''$, *respectively. Then for any set of selectable vectors with the membership function* $\lambda_Y^C(\cdot)$ *we have the inequalities*

$$\lambda_Y^C(y) \leqq \lambda_Y^M(y) \leqq \lambda_Y^P(y) \quad \textit{for all } y \in Y,$$

where $\lambda_Y^P(y)$ *is the membership function of the Pareto set and* $\lambda_Y^M(y)$ *is the membership function defined by*

$$\lambda_Y^M(y) = 1 - \sup_{z \in Y} \varsigma(z, y) \quad \textit{for all } y \in Y,$$

$$\varsigma(z,y) = \begin{cases} 1, & \textit{if } z - y \in R^m_+, \\ \mu_1, & \textit{if } \bar{z} - \bar{y} \in R^m_+,\ z - y \notin R^m_+, \\ \mu_2, & \textit{if } \hat{z} - \hat{y} \in R^m_+,\ \bar{z} - \bar{y} \notin R^m_+,\ z - y \notin R^m_+, \\ 0, & \textit{in the rest cases}, \end{cases} \quad \textit{for all } y, z, y \neq z,$$

and

$$\bar{y} = \left(y_1, \ldots, y_{k-1}, w_k' y_i + w_i' y_k, y_{k+1}, \ldots, y_m\right),$$
$$\hat{y} = \left(y_1, \ldots, y_{k-1}, w_k' w_j'' y_i + w_i' w_k'' y_j + w_i' w_j'' y_k, y_{k+1}, \ldots, y_m\right),$$
$$\bar{z} = \left(z_1, \ldots, z_{k-1}, w_k' z_i + w_i' z_k, z_{k+1}, \ldots, z_m\right),$$
$$\hat{z} = \left(z_1, \ldots, z_{k-1}, w_k' w_j'' z_i + w_i' w_k'' z_j + w_i' w_j'' z_k, z_{k+1}, \ldots, z_m\right).$$

The proof of this theorem is similar to that of the previous one, and we omit it.

7.4.3 Fuzzy Cyclic Information Quanta and Their Consistency

Definition 7.3 We say that there is given *fuzzy cyclic information* with the degrees of confidence $\mu_1, \ldots, \mu_k \in (0, 1]$ and positive parameters $w_{i_1}^{(1)}, w_{i_2}^{(1)}, w_{i_2}^{(2)}, w_{i_3}^{(2)}, \ldots,$ $w_{i_{k-1}}^{(k-1)}, w_{i_k}^{(k)}, w_{i_1}^{(k)}$ if the vectors $y^{(1)}, \ldots, y^{(k)} \in R^m$ defined by

$$\begin{array}{ll} y_{i_1}^{(1)} = w_{i_1}^{(1)}, y_{i_2}^{(1)} = -w_{i_2}^{(1)}, y_s^{(1)} = 0 & \text{for all } s \in I \backslash \{i_1, i_2\}, \\ y_{i_2}^{(2)} = w_{i_2}^{(2)}, y_{i_3}^{(2)} = -w_{i_3}^{(2)}, y_s^{(2)} = 0 & \text{for all } s \in I \backslash \{i_2, i_3\}, \\ \cdots\cdots\cdots\cdots\cdots\cdots\cdots\cdots\cdots\cdots & \\ y_{i_k}^{(k)} = w_{i_k}^{(k)}, y_{i_1}^{(k)} = -w_{i_1}^{(k)}, y_s^{(k)} = 0 & \text{for all } s \in I \backslash \{i_1, i_k\}, \end{array}$$

satisfy the equalities $\mu(y^{(1)}, 0_m) = \mu_1, \ldots, \mu(y^{(k)}, 0_m)) = \mu_k$.

As cyclic information is specified using a collection of quanta, it is necessary to consider the issue of consistency. The definition of a consistent collection of vectors has been given in Sect. 7.2.2. For numbers $\mu_1, \ldots, \mu_k$, also recall the permutation

$$\mu_{11}, \ldots, \mu_{1k_1}, \mu_{2k_2}, \ldots, \mu_{2k_2}, \ldots \mu_{l1}, \ldots, \mu_{lk_l},$$

having the property

$$1 \geqq \mu_{11} = \ldots = \mu_{1k_1} > \mu_{21} = \ldots = \mu_{2k_2} > \ldots > \mu_{l1} = \ldots = \mu_{lk_l} > 0,$$

where $k = k_1 + \ldots + k_l$, $1 \geqq l \geqq k$. In the sequel, we will use the crisp cones K_h, $h \in \{1, \ldots, l\}$, generated by the unit vectors e^1, ..., e^m of space R^m together with the vectors $y^{(i)}$, $i \in \{1, \ldots, k\}$, such that the corresponding numbers μ_i satisfy the inequality $\mu_i \geqq \mu_{h1}$. These cones are nested: $K_1 \subset K_2 \subset \ldots \subset K_l$.

Consider the matrix

$$W = \begin{pmatrix} w_{i_1}^{(1)} & 0 & \ldots & 0 & -w_{i_1}^{(k)} \\ w_{i_2}^{(1)} & w_{i_2}^{(2)} & \ldots & 0 & 0 \\ \vdots & \vdots & \vdots & \vdots & \vdots \\ 0 & 0 & \ldots & w_{i_{k-1}}^{(k-1)} & 0 \\ 0 & 0 & \ldots & -w_{i_k}^{(k-1)} & w_{ik}^{(k)} \end{pmatrix}.$$

Demonstrate that *the fuzzy cyclic information introduced by Definition* 7.3 *is consistent if and only if the matrix W has the positive determinant, i.e.,* $|W| > 0$.

☐ Necessity. Let the fuzzy cyclic information be consistent. According to Theorem 7.2, for the vectors forming this collection of cyclic information quanta, the system of linear Eq. (7.3) has no *N*-solution. In this case, $|W| > 0$ by Theorem 5.6.

Sufficiency. Let $|W| > 0$. By Theorem 5.6, the corresponding system of homogeneous linear Eqs. (7.3) has no *N*-solution. Prove that each cone K_h, $h \in \{1, \ldots, l-1\}$, does not contain any vectors $y^{(j)}$, $j = 1, 2, \ldots, k$, such that the corresponding number μ_j obeys the inequality $\mu_j < \mu_{h1}$. Conjecture the opposite, i.e., there exist numbers $\bar{h} \in \{1, \ldots, l-1\}$ and $\bar{j} \in \{1, \ldots, k\}$ such that $\mu_{\bar{j}} < \mu_{\bar{h}1}$, and the cone $K_{\bar{h}}$ contains the vector $y^{(\bar{j})}$.

Assume that $\bar{j} \in \{1, \ldots, k-1\}$ (the case $\bar{j} = k$ is treated by analogy). On the strength of the aforesaid, the vector $y^{(\bar{j})}$ can be expressed as the linear nonnegative combination

$$y^{(\bar{j})} = \sum_{i=1}^{m} \bar{\lambda}_i e^i + \sum_{j:\, \mu_j \geqq \mu_{\bar{h}1}} \bar{\xi}_j y^{(j)}. \tag{7.11}$$

Here two cases are possible, namely, $\mu_{\bar{j}+1} \geqq \mu_{\bar{h}1}$ or $\mu_{\bar{j}+1} < \mu_{\bar{h}1}$. Then the components of the vectors $y^{(\bar{j})}$ and $y^{(\bar{j}+1)}$ have the following form that corresponds to these cases:

$$y_p^{(\bar{j})} = w_p^{(\bar{j})},\ y_q^{(\bar{j})} = -w_q^{(\bar{j})},\ y_s^{(\bar{j})} = 0 \qquad \text{for all } s \in I \backslash \{p,q\},$$
$$y_q^{(\bar{j}+1)} = w_q^{(\bar{j}+1)},\ y_t^{(\bar{j}+1)} = -w_t^{(\bar{j}+1)},\ y_s^{(\bar{j}+1)} = 0 \quad \text{for all } s \in I \backslash \{q,t\},$$

where p, q, and t are some pairwise different numbers of the criteria from the collection $\{i_1, i_2, \ldots, i_k\}$. In the first and second cases above, component q of the vector equality (7.11) is defined by

$$-w_q^{(\bar{j})} = \bar{\lambda}_q + \bar{\xi}_{\bar{j}+1} w_q^{(\bar{j}+1)} \text{ and } - w_q^{(\bar{j})} = \bar{\lambda}_q,$$

respectively. Both of these equalities fails under any positive numbers $\bar{\lambda}_q$ and $\bar{\xi}_{\bar{j}+1}$. Hence, the hypothesis $y^{(\bar{j})} \in K_{\bar{h}}$ is wrong and, by Theorem 7.2, the fuzzy cyclic information is consistent. ■

7.4.4 Pareto Set Reduction Via Elementary Fuzzy Cyclic Information Quantum

Consider given fuzzy cyclic information quantum with the degrees of confidence $\mu_1, \mu_2, \mu_3 \in (0,\ 1]$. According to Definition 7.2, this means that the vectors $y^{(1)}, y^{(2)}, y^{(3)} \in R^m$ with the components

$$y_i^{(1)} = w_i^{(1)},\quad y_j^{(1)} = -w_j^{(1)},\quad y_s^{(1)} = 0 \quad \text{for all } s \in I \backslash \{i,j\},$$
$$y_j^{(2)} = w_j^{(2)},\quad y_l^{(2)} = -w_l^{(2)},\quad y_s^{(2)} = 0 \quad \text{for all } s \in I \backslash \{j,l\},$$
$$y_l^{(3)} = w_l^{(3)},\quad y_i^{(3)} = -w_i^{(3)},\quad y_s^{(3)} = 0 \quad \text{for all } s \in I \backslash \{l,i\},$$

satisfy $\mu(y^{(1)}, 0_m) = \mu_1$, $\mu(y^{(2)}, 0_m) = \mu_2$, $\mu(y^{(3)}, 0_m) = \mu_3$. Suppose that this fuzzy cyclic information is consistent, i.e., $|W| > 0$.

Order the numbers μ_1, μ_2, μ_3 so that $\tilde{\mu}_1 \geqq \tilde{\mu}_2 \geqq \tilde{\mu}_3$, where $\tilde{\mu}_1, \tilde{\mu}_2, \tilde{\mu}_3$ is some permutation of μ_1, μ_2, μ_3.

Analyze the following situation.

(I) $\mu_1 \geqq \mu_2 \geqq \mu_3$. These inequalities correspond to the ordered collection of vectors $y^{(1)}$, $y^{(2)}$, $y^{(3)}$ that forms the "chain" of quanta: criterion i is more important than criterion j, criterion j is more important than criterion l, and criterion l is more important than criterion i.

Situation (I) also covers the cases $\mu_3 \geqq \mu_1 \geqq \mu_2$ and $\mu_2 \geqq \mu_3 \geqq \mu_1$. For instance, the first case corresponds to the vectors $y^{(3)}$, $y^{(1)}$, $y^{(2)}$ forming the "chain" of quanta described by situation (I) with index l replaced by index i, index i by index j, and index j by index l.

Now, consider another situation.

(II) $\mu_1 > \mu_3 > \mu_2$. The corresponding ordered collection of vectors $y^{(1)}$, $y^{(3)}$, $y^{(2)}$ forms the following "chain" of quanta: criterion i is more important than criterion j, criterion l is more important than criterion i, and criterion j is more important than criterion l. Again, with an appropriate change of indexes, the same situation describes the inequalities $\mu_2 > \mu_1 > \mu_3$ and $\mu_3 > \mu_2 > \mu_1$.

Since there exist just 6 permutations of three numbers, situations (I) and (II) exhaust all possible inequalities for μ_1, μ_2, and μ_3. Situation (I) involves nonstrict inequalities, while situation (II) the strict ones.

Consider situation (I). Without loss of generality, let $\mu_1 \geqq \mu_2 \geqq \mu_3$ (otherwise, simply renumber the criteria). Define the membership function

$$\lambda_Y^M(y) = 1 - \sup_{z \in Y} \varsigma(z, y) \quad \text{for all } y \in Y, \tag{7.12}$$

where

$$\varsigma(z,y) = \begin{cases} 1, & \text{if } z - y \in R_+^m, \\ \mu_1, & \text{if } \bar{z} - \bar{y} \in R_+^m,\ z - y \notin R_+^m, \\ \mu_2, & \text{if } \tilde{z} - \tilde{y} \in R_+^m,\ \bar{z} - \bar{y} \notin R_+^m, \\ \mu_3, & \text{if } \hat{z} - \hat{y} \in R_+^m,\ \tilde{z} - \tilde{y} \notin R_+^m, \\ 0, & \text{in the rest cases,} \end{cases} \quad \text{for all } y, z \in Y, y \neq z, \tag{7.13}$$

and the vectors $\bar{a}$, $\tilde{a}$, $\hat{a}$, $a = (a_1, a_2, \ldots, a_m) \in \{y, z\}$ have the form

$$\begin{aligned} \bar{a} &= a + (w_j^{(1)} a_i + (w_i^{(1)} - 1) a_j) e^j, \\ \tilde{a} &= a + (w_j^{(1)} a_i + (w_i^{(1)} - 1) a_j) e^j + (w_l^{(2)} w_j^{(1)} a_i + w_l^{(2)} w_i^{(1)} a_j + (w_j^{(2)} w_i^{(1)} - 1) a_l) e^l, \\ \hat{a} &= a + ((w_j^{(2)} w_l^{(3)} - 1) a_i + w_l^{(2)} w_i^{(3)} a_j + w_j^{(2)} w_i^{(3)} a_l) e^i + (w_j^{(1)} w_l^{(3)} a_i + (w_i^{(1)} w_l^{(3)} - 1) a_j \\ &\quad + w_j^{(1)} w_i^{(3)} a_l) e^j + (w_l^{(2)} w_j^{(1)} a_i + w_l^{(2)} w_i^{(1)} a_j + (w_j^{(2)} w_i^{(1)} - 1) a_l) e^l. \end{aligned}$$

Theorem 7.8 *Assume that there is given consistent cyclic information quantum described by situation (I). Then for any fuzzy set of selectable vectors $C(Y)$ with the membership function $\lambda_Y^C(\cdot)$ we have the inequalities*

$$\lambda_Y^C(y) \leqq \lambda_Y^M(y) \leqq \lambda_Y^P(y) \quad \text{for all } y \in Y,$$

where the function $\lambda_Y^M(\cdot)$ is defined by (7.12)–(7.13).

□ The specification of consistent fuzzy cyclic information means that $\mu(y^{(1)}, 0_m) = \mu_1$, $\mu(y^{(2)}, 0_m) = \mu_2$, and $\mu(y^{(3)}, 0_m) = \mu_3$ for vectors $y^{(1)}$, $y^{(2)}$, and $y^{(3)}$.

Consider the three crisp cones K_1, K_2, and K_3 such that $K_1 \subset K_2 \subset K_3$, see above. The vector $y^{(i)}$ belongs to the cone K_i for all $i \in \{1, 2, 3\}$.

For each cone K_i, $i = 1, 2, 3$, introduce a fuzzy cone M_i such that supp $M_i = K_i$ and all its elements belonging to the nonnegative orthant (except the origin) have the unit grade of membership, while the rest elements have the grade of membership μ_i.

Let M be a fuzzy cone with the membership function $\eta(\cdot)$ and the support formed by the nonnegative orthant R^m_+ without the origin and by the vectors $y^{(1)}$, $y^{(2)}$, and $y^{(3)}$, where

$$\eta(y) = \begin{cases} 1, & \text{if } y \in R^m_+, \\ \mu_1, & \text{if } y \in K_1,\ y \notin R^m_+, \\ \mu_2, & \text{if } y \in K_2,\ \notin K_1, \\ \mu_3, & \text{if } y \in K_3,\ y \notin K_2, \\ 0, & \text{in the rest cases.} \end{cases}$$

The cone M is the union of the cones M_1, M_2, and M_3; moreover, it is acute and convex due to the consistency of the given information.

Based on the proof of Theorem 5.7, the set of vectors making the support of the cone M (i.e., K_3) coincides with the set of all nonzero solutions to the system of inequalities

$$\begin{gathered} w_j^{(2)} w_l^{(3)} y_i + w_i^{(3)} w_l^{(2)} y_j + w_i^{(3)} w_j^{(2)} y_l \geqq 0, \\ w_j^{(1)} w_l^{(3)} y_i + w_i^{(1)} w_l^{(3)} y_j + w_i^{(3)} w_j^{(1)} y_l \geqq 0, \\ w_j^{(1)} w_l^{(2)} y_i + w_i^{(1)} w_l^{(2)} y_j + w_i^{(1)} w_j^{(2)} y_l \geqq 0, \\ y_s \geqq 0 \quad \text{for all } s \in I \backslash \{i, j, l\}. \end{gathered}$$

The inclusion $y \in \text{supp}\, M_2 = K_2$ is equivalent to the following system of homogeneous linear inequalities (where at least one inequality is strict):

$$\begin{gathered} w_j^{(1)} y_i + w_i^{(1)} y_j \geqq 0, \\ w_j^{(1)} w_l^{(2)} y_i + w_i^{(1)} w_l^{(2)} y_j + w_i^{(1)} w_j^{(2)} y_l \geqq 0, \\ y_s \geqq 0 \quad \text{for all } s \in I \backslash \{j, l\}. \end{gathered}$$

And the inclusion $y \in \text{supp}\, M_1 = K_1$ is equivalent to the system of inequalities

$$\begin{gathered} w_j^{(1)} y_i + w_i^{(1)} y_j \geqq 0, \\ y_s \geqq 0 \quad \text{for all } s \in I \backslash \{j\}, \end{gathered}$$

where at least one inequality is strict, see Theorem 2.5.

Consider the fuzzy cone relation with the cone M. Denote by $\psi(\cdot, \cdot)$ its membership function, which satisfies $\psi(z, y) = \eta(z - y)$ for all $z, y \in Y$. Obviously, this function is exactly $\varsigma(\cdot, \cdot)$ defined by (7.13). The fuzzy binary relation ζ is hence the cone relation with the cone M.

As follows from (7.12), the fuzzy set with the membership function $\lambda_Y^M(\cdot)$ makes the fuzzy set of nondominated vectors in terms of the relation ς. The nonnegative orthant R_+^m is contained in M by definition and, in addition, we have the inclusion $M \subset K$. The Pareto set therefore includes the fuzzy set of nondominated vectors in terms of the binary relation ς, which includes the fuzzy set of nondominated vectors in terms of the relation μ.

Let $\lambda_Y^N(\cdot)$ be the membership function of the fuzzy set of nondominated vectors in terms of the binary relation μ (the latter corresponds to the cone K). The inequality $\lambda_Y^C(y) \leqq \lambda_Y^N(y)$ holds for all $y \in Y$ by Lemma 7.1. This yields the desired inequalities $\lambda_Y^C(y) \leqq \lambda_Y^N(y) \leqq \lambda_Y^P(y)$ for all $y \in Y$. ■

According to Theorem 7.8, one has to solve four multicriteria problems with crisp preference relations in order to reduce the Pareto set using fuzzy cyclic information. First, it is necessary to find the Pareto set $P(Y)$ for the original multicriteria problem with respect to the initial vector function f and the set of feasible alternatives X; then assign $\lambda_Y^M(y) = 1$ for all vectors $y \in P(Y)$ and $\lambda_Y^M(y) = 0$ for all other vectors from the set Y. Second, it is necessary to solve the multicriteria problem with the new vector criterion $\bar{f}$ obtained from f by replacing f_j with the linear combination $w_j^{(1)} f_i + w_i^{(1)} f_j$. Denote $\bar{P}(Y) = f(P_{\bar{f}}(X))$ and assign the grade of membership $1 - \mu_1$ to the vectors belonging to the set $P(Y) \backslash \bar{P}(Y)$. Third, it is necessary to solve the multicriteria problem with the new vector criterion $\tilde{f}$ obtained from f by replacing f_j with the linear combination $w_j^{(1)} f_i + w_i^{(1)} f_j$ and f_i with the linear combination $w_l^{(2)} w_j^{(1)} f_i + w_l^{(2)} w_i^{(1)} f_j + w_j^{(2)} w_i^{(1)} f_l$. Denote by $\tilde{P}(Y)$ the image of the set of Pareto optimal alternatives $P_{\tilde{f}}(X)$ under the mapping f; then assign the grade of membership $1 - \mu_2$ to all vectors in the set $\bar{P}(Y) \backslash \tilde{P}(Y)$. And fourth, it is necessary to solve the multicriteria problem with the new vector criterion $\hat{f}$ obtained from f by replacing f_i, f_j, and f_l with the linear combinations $w_j^{(2)} w_l^{(3)} f_i + w_l^{(2)} w_i^{(3)} f_j + w_j^{(2)} w_i^{(3)} f_l$, $w_j^{(1)} w_l^{(3)} f_i + w_i^{(1)} w_l^{(3)} f_j + w_j^{(1)} w_i^{(3)} f_l$, and $w_l^{(2)} w_j^{(1)} f_i + w_l^{(2)} w_i^{(1)} f_j + w_j^{(2)} w_i^{(1)} f_l$, respectively. Again, denote $\hat{P}(Y) = f(P_{\hat{f}}(X))$ and then assign the grade of membership $1 - \mu_3$ to all vectors in the set $\tilde{P}(Y) \backslash \hat{P}(Y)$.

And so, by solving the four multicriteria problems above, we construct the desired membership function $\lambda_Y^M(\cdot)$ of the fuzzy set that represents the reduced Pareto set.

Remark 7.1 Let $\mu_1 = \mu_2$ under the hypotheses of Theorem 7.8. Then for all $y, z \in Y, y \neq z$, the fuzzy cone relation ζ is calculated by the formula

$$\varsigma(z, y) = \begin{cases} 1, & \text{if } z - y \in R_+^m, \\ \mu_1 = \mu_2, & \text{if } \tilde{z} - \tilde{y} \in R_+^m,\ z - y \notin R_+^m, \\ \mu_3, & \text{if } \hat{z} - \hat{y} \in R_+^m,\ \tilde{z} - \tilde{y} \notin R_+^m, \\ 0, & \text{in the rest cases.} \end{cases}$$

Remark 7.2 Let $\mu_2 = \mu_3$ under the hypotheses of Theorem 7.8. Then for all $y, z \in Y, y \neq z$, the fuzzy cone relation ζ is calculated by the formula

$$\varsigma(z,y) = \begin{cases} 1, & \text{if } z - y \in R^m_+, \\ \mu_1, & \text{if } \bar{z} - \bar{y} \in R^m_+,\ z - y \notin R^m_+, \\ \mu_2 = \mu_3, & \text{if } \hat{z} - \hat{y} \in R^m_+,\ \bar{z} - \bar{y} \notin R^m_+, \\ 0, & \text{in the rest cases.} \end{cases}$$

Remark 7.3 Let $\mu_1 = \mu_2 = \mu_3 = \mu^*$ under the hypotheses of Theorem 7.8. Then for all $y, z \in Y, y \neq z$, the fuzzy cone relation ζ is calculated by the formula

$$\varsigma(z,y) = \begin{cases} 1, & \text{if } z - y \in R^m_+, \\ \mu^*, & \text{if } \hat{z} - \hat{y} \in R^m_+,\ z - y \notin R^m_+, \\ 0, & \text{in the rest cases.} \end{cases}$$

Example 7.3 Consider an illustrative example where the set of feasible vectors Y consists of the six three-dimensional vectors

$$y^{(1)} = (4,2,1),\ y^{(2)} = (5,1,3),\ y^{(3)} = (1,0,1),$$
$$y^{(4)} = (3,4.5,3),\ y^{(5)} = (1,5,3),\ y^{(6)} = (2,4,4).$$

Assume that there is cyclic information about the DM fuzzy preference relation in the form of the vectors

$$v^{(1)} = (3,-1,0),\ v^{(2)} = (0,2,-3),\ v^{(3)} = (-2,0,4)$$

with the degrees of confidence 0.9, 0.7 and 0.3. In other words, we have

$$\mu(v^{(1)}, 0_3) = 0.9,\ \mu(v^{(2)}, 0_3) = 0.7,\ \mu(v^{(3)}, 0_3) = 0.3.$$

Obviously, this information is consistent due to the positive determinant $|W| = |v^{(1)}v^{(2)}v^{(3)}| = 18$. All vectors of the set Y (except $y^{(3)}$) are Pareto optimal. Hence, the membership function λ^P_Y of the Pareto set $P(Y)$ is

$$\lambda^P_Y(y^{(i)}) = 1 \quad \text{for } i = 1,2,4,5,6;\ \lambda^P_Y(y^{(3)}) = 0.$$

This case corresponds to situation (I) (0.9 > 0.7 > 0.3), and we will construct the "new" Pareto set with the membership function λ^M_Y using Theorem 7.8 based on the fuzzy cyclic information. Solve the second multicriteria problem with the vector criterion $\bar{f} = (\bar{f}_1, \bar{f}_2, \bar{f}_3)$, where $\bar{f}_1 = f_1$, $\bar{f}_2 = f_1 + 3f_2$, $\bar{f}_3 = f_3$. Using these relationships, form the set of feasible vectors $\bar{Y}$ as follows:

$$\bar{y}^{(1)} = (4, 10, 1),\ \bar{y}^{(2)} = (5, 8, 3),\ \bar{y}^{(3)} = (1, 1, 1),$$
$$\bar{y}^{(4)} = (3, 16.5, 3),\ \bar{y}^{(5)} = (1, 16, 3),\ \bar{y}^{(6)} = (2, 14, 4).$$

Here the first, second, fourth and sixth vectors are Pareto optimal. At this stage, the membership function λ_Y^M is given by

$$\lambda_Y^M(y^{(i)}) = 1 \quad \text{for } i = 1, 2, 4, 6, \quad \lambda_Y^M(y^{(5)}) = 0.1,\ \lambda_Y^M(y^{(3)}) = 0.$$

Now, solve the third multicriteria problem with the vector criterion $\tilde{f} = (\tilde{f}_1, \tilde{f}_2, \tilde{f}_3)$, where $\tilde{f}_1 = f_1$, $\tilde{f}_2 = f_1 + 3f_2$, $\tilde{f}_3 = 3f_1 + 9f_2 + 6f_3$, and with the corresponding set of feasible vectors $\tilde{Y}$:

$$\tilde{y}^{(1)} = (4, 10, 36),\ \tilde{y}^{(2)} = (5, 8, 42),\ \tilde{y}^{(3)} = (1, 1, 9),$$
$$\tilde{y}^{(4)} = (3, 16.5, 67.5),\ \tilde{y}^{(5)} = (1, 16, 66),\ \tilde{y}^{(6)} = (2, 14, 66).$$

The Pareto optimal vectors are $\tilde{y}^{(1)}$, $\tilde{y}^{(2)}$, and $\tilde{y}^{(4)}$; hence,

$$\lambda_Y^M(y^{(i)}) = 1 \quad \text{for } i = 1, 2, 4,\ \lambda_Y^M(y^{(6)}) = 0.3,\ \lambda_Y^M(y^{(5)}) = 0.1,\ \lambda_Y^M(y^{(3)}) = 0.$$

Finally, solve the fourth multicriteria problem, where the vector criterion $\hat{f} = (\hat{f}_1, \hat{f}_2, \hat{f}_3)$ consists of the functions $\hat{f}_1 = 8f_1 + 6f_2 + 4f_3$, $\hat{f}_2 = 4f_1 + 12f_2 + 2f_3$, and $\hat{f}_3 = 3f_1 + 9f_2 + 6f_3$. The set of feasible vectors $\hat{Y}$ is constructed in the following way:

$$\hat{y}^{(1)} = (48, 42, 36),\ \hat{y}^{(2)} = (58, 38, 42),\ \hat{y}^{(3)} = (12, 6, 9),$$
$$\hat{y}^{(4)} = (63, 72, 67.5),\ \hat{y}^{(5)} = (50, 70, 66),\ \hat{y}^{(6)} = (56, 64, 66).$$

Since the vector $\hat{y}^{(4)}$ dominates the others in terms of the Pareto relation, the membership function λ_Y^M acquires the form

$$\lambda_Y^M(y^{(4)}) = 1,\ \lambda_Y^M(y^{(1)}) = \lambda_Y^M(y^{(2)}) = 0.7,$$
$$\lambda_Y^M(y^{(6)}) = 0.3,\ \lambda_Y^M(y^{(5)}) = 0.1,\ \lambda_Y^M(y^{(3)}) = 0.$$

Thus, using the cyclic information about the fuzzy preference relation, we have reduced the initial Pareto set to the fuzzy set with the membership function λ_Y^M.

Let us study situation (II). Without loss of generality, suppose that $\mu_1 > \mu_3 > \mu_2$ (i.e., $\tilde{\mu}_1 = \mu_1$, $\tilde{\mu}_2 = \mu_3$, $\tilde{\mu}_3 = \mu_2$); otherwise, just renumber the criteria appropriately.

As before, consider the membership function $\lambda_Y^M(\cdot)$ defined by (7.12). Here for all $y, z \in Y, y \neq z$, the fuzzy relation ζ has the membership function

$$\varsigma(z,y) = \begin{cases} 1, & \text{if } z - y \in R^m_+, \\ \mu_1, & \text{if } \bar{z} - \bar{y} \in R^m_+,\ z - y \notin R^m_+, \\ \mu_3, & \text{if } \tilde{z} - \tilde{y} \in R^m_+,\ \bar{z} - \bar{y} \notin R^m_+, \\ \mu_2, & \text{if } \hat{z} - \hat{y} \in R^m_+,\ \tilde{z} - \tilde{y} \notin R^m_+, \\ 0, & \text{in the rest cases.} \end{cases} \tag{7.14}$$

while the vectors $\bar{b}$, $\tilde{b}$, $\hat{b}$, $b = (b_1, \ldots, b_m) \in \{y, z\}$, are such that $\bar{b} = \bar{a}$, $\hat{b} = \hat{a}$, and

$$\tilde{b} = b + (w_i^{(3)} b_l + (w_l^{(3)} - 1) b_i) e^i + (w_j^{(1)} w_l^{(3)} b_i + (w_i^{(1)} w_l^{(3)} - 1) b_j + w_j^{(1)} w_i^{(3)} b_l) e^j.$$

Theorem 7.9 *Assume that there is given consistent fuzzy cyclic information described by situation* (II). *Then for any membership function $\lambda_Y^M(\cdot)$ of the fuzzy set of selectable vectors* $C(Y)$, *we have the inequalities*

$\lambda_Y^C(y) \leqq \lambda_Y^M(y) \leqq \lambda_Y^P(y)$ *for all* $y \in Y$,

where the function $\lambda_Y^M(\cdot)$ is defined by (7.12) and (7.14).

The proof of this result resembles that of Theorem 7.9 associated with situation (I), except that

(a) the crisp cone K_2 is generated by the nonnegative orthant and the vectors $y^{(1)}, y^{(3)}$ and
(b) the support of the fuzzy cone M_2 (the cone K_2) is the set of all nonzero solutions to the following system of homogeneous inequalities (with at least one strict inequality):

$$\begin{gathered} w_i^{(3)} y_l + w_l^{(3)} y_i \geqq 0, \\ w_j^{(1)} w_l^{(3)} y_i + w_i^{(1)} w_l^{(3)} y_j + w_i^{(3)} w_j^{(1)} y_l \geqq 0, \\ y_s \geqq 0 \quad \text{for all } s \in I \backslash \{i, j\}. \end{gathered}$$

Chapter 8
Decision-Making Based on Information Quanta: Methodology and Practice

This chapter considers in brief the aspects of decision-making by humans and then presents the axiomatic approach to Pareto set (domain of compromise) reduction based on information quanta about the DM's preference relation. The corresponding theoretical background can be found in the previous chapters, and here we describe the axiomatic approach without mathematical details, as well as give some recommendations on usage. In addition, possible ways to combine this approach with some multicriteria scalarization methods and some potential extensions are discussed.

8.1 How Do Humans Make Their Decisions?

8.1.1 Mental Components of Decision-Making Process

Decision-making process includes three phases, namely, search, choice and implementation of decisions.

Decision-making is an act of volition that forms a sequence of actions towards goal attainment by transforming initial information under uncertainty. The main stages of decision-making process are situational analysis using available information and the decision-making procedure itself, i.e., the formation and comparison of alternatives, the choice of an appropriate alternative and the development of an action plan.

On the one hand, decision-making may represent a special form of mental activity (e.g., in management) and, on the other, as a stage of thinking in problem solving. This notion has a very wide range of application. Throughout the book, we understand decision-making as a special process of human activity intended for choosing a best alternative (a best action).

© Springer International Publishing AG 2018

V.D. Noghin, *Reduction of the Pareto Set*, Studies in Systems, Decision and Control 126, https://doi.org/10.1007/978-3-319-67873-3_8

Decision-making process is functionally supported by intelligence based on the joint activity of memory, attention and thinking.

Memory constructs a bridge between the past of a subject and its present and future, representing special processes for the organization and retention of accumulated experience; these processes allow to use repeatedly the existing experience in human activity or even to recur to consciousness. Memory underlies any mental phenomena. As a matter of fact, an individual together with its relations, skills, habits, hopes, desires and claims exists owing to its memory.

Memory has different types depending on retention period, namely, momentary or sensory memory (retention period is less than 1 s), short-term memory (retention period is up to 30 s), working memory (retention period is up to several minutes), and long-term memory (retention period is from several hours to decades). Psychologists believe that decision-making is mostly associated with working memory, since information retention for further decision-making is a typical scenario for this memory. Working memory has a close connection with long-term memory and is based on different methods and techniques of remembering developed in other kinds of activity. By-turn, long-term memory employs the methods and techniques of remembering that are established within working memory. These types of memory are closely connected in the sense of information flows: working memory uses some information stored in long-term memory and, at the same time, transmits a certain portion of new information to long-term memory.

Interestingly, the working memory of a human has a limited capacity. This fact is known as Miller's law, in the honor of psychologist George A. Miller who published in 1956 his famous paper [28] on the magical number 7 ± 2. He claimed that the information-processing capacity of young adults is around seven simultaneous elements, which were called "chunks", regardless whether the elements are digits, letters, words, or other units. Later research revealed that this number depends on the category of chunks used (e.g., span may be around seven for digits, six for letters, and five for words), and even on features of the chunks within a category.

The statement of the memory problem was appreciably affected by the analogy between data processing by humans and the structural blocks of computers. Note however that the functional structure of human memory demonstrates higher flexibility against that of a computer.

The next component of intelligence concerns attention, which is often comprehended as the concentration of the subject's activity on an ideal or real object (an item, event, image, idea, etc.) at a given moment. Attention makes the dynamic side of consciousness that characterizes the degree of orientation to (focus on) an object for its adequate reflection during the implementation of a definite activity (e.g., decision-making). Attention allows an individual to direct consciousness towards objects it perceives in the course of certain activity. A human does a job faster and with higher quality owing to focused attention. On the other hand, improper attention complicates the perception of new information and learning. As is well-known, inattentive behavior has a pernicious effect, e.g., on calculations: merely a single mistake yields an incorrect result.

Psychology defines thinking as the cognitive activity of a human that ensures information organization and processing; it forms the analysis, synthesis and generalization of conditions and requirements of a problem and solution methods. Owing to developed thinking, a human can surmount the space limitation of perception, directing its thoughts towards the immense ranges of the macro- and microworld. Note that the time limitation of perception is also eliminated, as thinking gives free motion along the time axis from the ancient periods to the uncertain future.

Thinking is activated while a human solves a new problem, which is topical and has no ready-made solutions; and a powerful motive stimulates it to find a way out. The origin and realization of a new problem form the direct spur that initiates mental process. The next stage is often connected with the delay of impulsive reactions. This delay creates a necessary pause for the analysis of the problem and its conditions, as well as for the extraction of essential components and their correlation with each other. The key stage of thinking is to choose a certain alternative and design a general scheme of solution.

Thinking includes voluntary and involuntary components. Among the examples of involuntary components, we mention associations causing uncontrolled relations that may define stereotyped reasoning (on the one hand) or yield the original and fruitful ideas and hypotheses in the context of the problem (on the other hand). Thinking has the unity of conscious and unconscious as its characteristic feature. It should be remembered that a major role in mental activity is played by emotions that guide problem-solving.

There exist several types of thinking, namely, visual image thinking, verbal image thinking, verbal logical thinking, etc. As generally accepted, verbal logical thinking is the latest product in the development of human thinking and transition from visual thinking to abstract thinking makes a line of this development. Moreover, psychologists identify the following (opposite) types of thinking: theoretical and practical (empirical), logical (analytical) and intuitive, realistic and autistic (cescape from reality into internal experience), and so others.

8.1.2 Decision-Making Strategy of Humans in Multicriteria Environment

In many situations the result of choice cannot be assessed using one scale, e.g., in terms of money or time. Though, the well-known idiom "time's money" states (at least, in theory) that time can be converted into money. In other words, one scale is in principle reducible to the other. But the idiom "man shall not live by bread alone" acts as a counterbalance here. We believe that the latter confirms the multicriteria character of the environment humans live in, as well as the irreducibility of the spiritual to the material. Hence, many aspects connected with a human cannot be expressed in a single scale.

The above idioms can be considered as a concentrated expression of the two fundamentally different positions reflecting the opposite viewpoints on the subject. According to the first viewpoint, there exists a uniform index or criterion to measure all qualities. And the second one claims that such an index does not exist in principle. It seems that none of these positions can be logically proved or rejected at the conceptual level, and therefore both have right to exist. But the second one ("man shall not live by bread alone") is more realistic and viable, since the abstract fact that everything can be expressed in a uniform scale is of little assistance in practical decision-making: this viewpoint requires implementation. In other words, it is necessary to master such reduction to a uniform scale (to perform the *scalarization* of the multicriteria problem in terms of multicriteria optimization), and this scalarization forms nothing but a definite stage of solution for the initial multicriteria problem.

The multicriteria problems represent an extremely difficult class of problems arising in the intellectual activity of humans. The presence of multiple criteria increases the load on the limited-capacity working memory of a human, making the problem more uncertain and requiring focused attention and, in many cases, unconventional thinking.

Nowadays, there is no clear picture how (using which mechanisms) a human performs choice in a multicriteria environment. There exist only certain approaches and proposals how to deal with these complex issues. Note that they often contradict each other and do not exhaust all possible ways of choice. It is considered that a behavioral trait of an individual that faces a choice problem consists in decomposing the initial problem into a set of simpler intermediate subproblems.

Let us discuss the elementary case where a human chooses between two feasible alternatives using multiple criteria. Then the behavioral strategies (see [54]) can be divided into two classes, namely,

- the compensatory strategy,
- the noncompensatory strategy (elimination).

The compensatory strategy corresponds to a line of behavior under which small values of one criterion (or a group of criteria) are balanced (compensated) by large values of another criterion (or another group of criteria). For instance, buying a car is a typical example of choice with the compensatory strategy: low fuel economy can be compensated by stylish appearance or brand image. Another example is buying an imperfect-layout house at a slightly overestimated price but in an attractive urban district near a public green space in close proximity to the place of employment.

The noncompensatory strategy is to remove from the list of existing feasible alternatives all ones that do not satisfy one or several criteria. While buying a car or house, an individual with the noncompensatory strategy directly eliminates all alternatives that go beyond its financial possibilities. The situation where the buyer's attention is focused on the cars with automatic transmission only (i.e., all cars with manual transmission are eliminated from consideration) gives another common example of noncompensatory strategy usage.

Numerous experiments show that humans do not adhere to a single line of behavior while solving multicriteria problems with more than two feasible alternatives. As a rule, humans adopt some combinations of these strategies. The decision-making mechanisms can be divided into two large classes, namely, exact (analytical, logical) and heuristic (approximate, intuitive) mechanisms, depending on the guarantee of their result. The mechanisms of the first class provide a clear description for the decision-making problems in which their application surely yields positive results (or at least eliminate all unacceptable decisions). And the heuristic mechanisms may give different results in terms of adequacy for different decision-making problems. Note that it is impossible to identify precisely the two subgroups of the decision-making problems where a given heuristic mechanism works well and does not so.

Many researchers reckon the decision-making mechanisms based on a mathematical apparatus as the exact mechanisms and methods. However, this approach might not be admitted, as the usage of a mathematical language to express a statement does not mean that the latter is exact. Moreover, humans without necessary background on mathematical subtleties would harbor an illusion that such mechanisms and methods have high accuracy and reliability.

8.2 Methodology of Axiomatic Approach Application for Pareto Set Reduction

8.2.1 Mathematical Modeling

In the simplified form decision-making process can be illustrated by the following scheme, see Fig. 8.1.

An appropriate decision (or decisions) is chosen by *the decision-maker* (DM). The latter also bears full responsibility for the decision. The solution of the multicriteria choice problem is called *the set of feasible alternatives* and denoted by $C(X)$. In some real problems this set represents a singleton. However, in common situations the set of feasible alternatives may include several (or even infinitely many) elements. For instance, the number of chosen candidates must coincide with the number of open vacancies.

The basic components of the multicriteria choice problem are *the set of feasible alternatives* X, *a vector criterion* $f = (f_1, f_2, \ldots, f_m)$ and *a preference relation* $\succ_X$ that guides the DM's choice.

To solve a specific choice problem, first one has to construct its mathematical model. In other words, it is necessary to form the set of feasible alternatives, the vector criterion and the preference relation that provide the most complete and precise description of the real situation. The higher is the adequacy of the

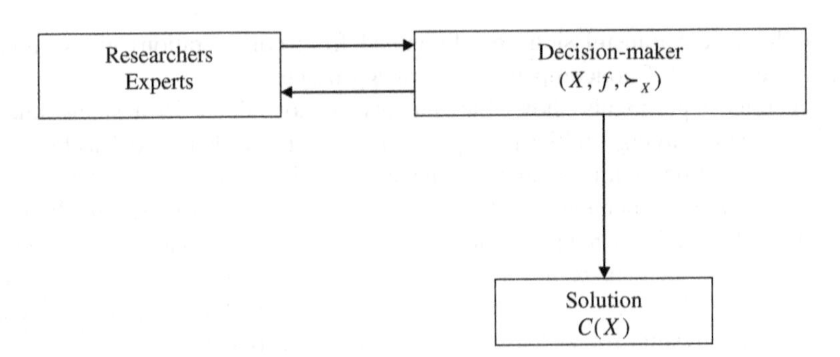

Fig. 8.1 Decision-making process

mathematical model to the real problem, the better is the chance to obtain the actually optimal solution.

The mathematical model is constructed by the DM together with researchers (specialists in the field of decision-making) and experts (specialists in the subject field of the initial problem). As a rule, exactly the joint and intensive efforts of these individuals yield a satisfactory mathematical model, which describes the real situation in adequate terms on the one hand and gives the best solution in reasonable time on the other. The first stage that is used to construct the mathematical model (the formalization stage) cannot be preset. Here many things depend on the experience and intuition of all participants.

The set of feasible alternatives can be finite, but it may also happen that this set consists of infinitely many elements. A finite set is often defined by enumerating all its elements. In the case of an infinite set, there exist different ways to do it (e.g., by the solution set of certain system of equations or inequalities). Further solution of the choice problem considerably depends on how the set of feasible alternatives is defined. Some definitions may appear inconvenient for the subsequent handling of sets. Here the final word rests with the specialist in decision-making.

Now, let us discuss the criteria. All functions $f_1, f_2, \ldots, f_m$ in the multicriteria choice problem must be numerical and the DM must be interested in the maximization of each this function. Imagine that the values of one or several criteria are measured in a qualitative scale. The experience shows that in such cases it is possible to pass to numerical values by introducing, e.g., a numerical rating scale. For instance, in Russia the academic performance of schoolchildren and students is assessed using the four-mark grade system (2, 3, 4, 5). Similar scales exist for assessing the performance of gymnasts and figure skaters in professional sports. Many examples of quantitative scales for measuring qualitative characteristics can be found in psychology. In this context we also refer to T. Saaty's 9-point scale [58].

If the DM seeks to minimize (not maximize) a certain criterion, then the latter is incorporated in the mathematical model with minus sign. Such a widespread technique transforms minimization into maximization. Note that, just like functions, criteria can be defined in different ways. Some situations require criteria with useful mathematical properties (e.g., continuity, differentiability, concavity or convexity). Again, here the DM needs a piece of advice from the specialist in decision-making.

The third component of the multicriteria choice problem–the preference relation–is most difficult to formalize. As a rule, it appears impossible to construct completely the preference relation that guides the DM's choice. One may acquire merely some fragmentary knowledge about this relation. Such knowledge must include information that the preference relation belongs to a definite class satisfying appropriate requirements. Recall that the original solution approach to the multicriteria choice problems that is suggested in the book implies that the DM's preference relation satisfies Axioms 1–4, which describe in a determinate sense the reasonable behavior of the subject in the course of decision-making.

According to *Axiom* 1, a certain alternative that is not chosen from a pair of alternatives belonging to the set of feasible alternatives will not be chosen from this set considered wholly. This requirement seems rather acceptable and so burdensome, but fails in some applications-relevant cases.

Next, *Axiom* 2 claims that the DM is in principle able to compare any vectors of the criterion space: for two arbitrary vectors $y, y' \in R^m$, only one of the following situations can be implemented:

- y is preferable to y', which is denoted by $y \succ y'$ (among these two vectors the DM chooses the first vector and does not choose the second one);
- y' is preferable to y, which is denoted by $y' \succ y$ (among these two vectors the DM chooses the second vector and does not choose the first one);
- Neither of the relationships $y \succ y'$ and $y' \succ y$ holds (among these two vectors the DM is unable to give preference to any).

By Axiom 2, the pairwise comparison results must have *transitivity*, i.e., for any triplet of vectors y, y', y'' satisfying the relationships $y \succ y'$ and $y' \succ y''$, we have $y \succ y''$. This property expresses the logical (reasonable) behavior of the DM during choice. However, despite the naturalness of this requirement, psychologists insist that humans do not always follow transitivity: while comparing three alternatives such that the first is better than the second and the second is better than the third, they may choose the third alternative between the first and third ones.

The whole essence of *Axiom* 3 is that the DM strives to maximize each of the criteria $f_1, f_2, \ldots, f_m$ under fixed values of the other criteria. Perhaps, it is clear that this requirement may also fail in some situation (e.g., if the DM is interested to maintain the value of a certain criterion within a definite range).

The last *Axiom* 4 dictates the following. For any two vectors y, y' of the criterion space R^m, the relationship $y \succ y'$ remains in force under (a) the simultaneous increase (or decrease) of all their components by the same number of times and (b) their addition with the same arbitrary vector from the criterion space. Properties (a) and (b) are called homogeneity and additivity, respectively, and their combination means the invariance property. Namely, let the relationship $y = (y_1, y_2, \ldots, y_m) \succ (y'_1, y'_2, \ldots, y'_m) = y'$ hold. Then by Axiom 4 we have the relationships $\alpha y = (\alpha y_1, \alpha y_2, \ldots, \alpha y_m) \succ (\alpha y'_1, \alpha y'_2, \ldots, \alpha y_m) = \alpha y'$ for an arbitrary positive number α and $y + c = (y_1 + c_1, y_2 + c_2, \ldots, y_m + c_m) \succ (y'_1 + c_1, y'_2 + c_2, \ldots, y'_m + c_m) = y' + c$ for any vector $c = (c_1, c_2, \ldots, c_m)$.

If the DM's preference relation does not satisfy at least one of the four axioms, the axiomatic approach below not necessarily yields the best result.

When these axioms are difficult to verify in a specific situation, it remains to hope that the axiomatic approach would not give an unsatisfactory solution.

8.2.2 *Elicitation of Information About DM's Preference Relation*

The main idea of the suggested approach is to use information about the DM's preference relation in the form of quanta in order to eliminate the (unacceptable) Pareto optimal alternatives. There exist at least two methods to acquire such information, namely,

- the analysis of the earlier actions undertaken by the DM,
- the direct questioning of the DM.

The first method proceeds from the knowledge of the DM's past behavior in similar choice problems with the criteria $f_1, f_2, \ldots, f_m$. If the DM has never faced such problems, the only way is to question the DM directly (in this case the second method leaves no option).

Prior to questioning, it is necessary to explain Definition 2.2 (or Definition 2.4) to the DM; recall that it covers the elementary case where criterion i (i.e., f_i) is more important than criterion j (i.e., f_j) with positive parameters w_i^* and w_j^*. This definition rests upon the compensatory principle discussed in the previous subsection: every time the DM is willing to sacrifice at most w_j^* units in terms of less important criterion j for gaining not less than w_i^* units in terms of more important criterion i under fixed values of the other criteria. In other words, the loss of maximum w_j^* units in terms of criterion j can be always compensated by obtaining minimum w_i^* units in terms of criterion i. The positive number

$$\theta_{ij} = \frac{w_j^*}{w_i^* + w_j^*} \quad (0 < \theta_{ij} < 1),$$

which expresses the ratio of the loss to the sum of the loss and gain, is called *the degree (coefficient) of compromise*.

If this coefficient is close to 1, then criterion i has very higher importance against criterion j, as the DM is willing to suffer a considerable loss in terms of a less important criterion for obtaining a relatively small increase in terms of a more important criterion. In the case $\theta_{ij} \approx 0$, the degree of compromise is small, since the

DM agrees to lose in terms of a less important criterion only under an appreciable gain in terms of a more important criterion.

However, the aforesaid should not be interpreted in absolute sense: the degree of compromise strongly depends on the measurement units of the criteria compared by importance (see Sect. 2.4). It may happen that two identical DMs (i.e., indistinguishable in terms of decision-making) solve the same problem using different degrees of compromise simply because they adopt different measurement units for the criteria compared by importance. And so, in a specific situation the degree of compromise depends on the measurement units of the criteria. The transition to other units (within the same scale!) often changes the degrees of compromise. For example, consider profits that are expressed in monetary units; then the degrees of compromise calculated by two identical DMs in different currencies (RUB and USD) would naturally differ.

Assume that in the course of questioning it turns out that the DM is willing to sacrifice a definite quantity in terms of criterion j for obtaining a certain gain in terms of criterion i. This situation indicates that criterion i has higher importance against criterion j according to Definition 2.4. It remains to define the degree of importance, i.e., to find the values of the parameters w_i^* and w_j^*. Here one should keep in mind an important aspect as follows: the higher is the degree of compromise $\theta_{ij} \in (0, 1)$, the more essential is the resulting information about the preference relation, and hence the greater reduction rate of the Pareto set (the domain of compromise) may be expected. Therefore, it is very desirable to clarify the maximum quantity w_j^* in terms of less important criterion j that the DM agrees to lose for gaining a fixed quantity w_i^* in terms of more important criterion i.

While eliciting the values of the parameters w_i^* and w_j^*, a convenient approach is to fix one of them (e.g., $w_i^* = 1$) and define only the other. In this case, the DM has to answer the following question: *what is your maximum admissible loss w_j^* in terms of less important criterion j for gaining minimum 1 unit in terms of more important criterion i?*

8.2.3 Sequential Reduction of Pareto Set

Let us describe the general sequential reduction scheme of the Pareto set based on a collection of information quanta. It involves the elimination (noncompensatory) strategy, see Sect. 8.1.2.

The first stage of this scheme is to elicit information about the preference relation in the form of quanta. The most widespread method lies in the direct questioning of the DM. This yields the pairs of more important and less important criteria f_i and f_j, as well as the corresponding parameters w_i^* and w_j^* that specify an elementary information quantum about the DM's preference relation.

The second stage runs without the DM. According to Theorem 2.5, it is necessary to replace the less important criterion f_j among the criteria $f_1, f_2, \ldots, f_m$ with the new criterion calculated by the simple formula $w_i^* f_j + w_j^* f_i$, and then to find the Pareto set (the so-called new Pareto set) in terms of the new vector criterion. This stage may cause some computational difficulties if the set of feasible alternatives is infinite. For a finite set of feasible vectors, the Pareto set can be constructed using the algorithm described in Sect. 1.6.

The Pareto set designed using the new vector criterion represents an upper estimate for the set of selectable alternatives. Simply speaking, further choice must be performed within this new Pareto set. And so, at *the third stage* the former set is shown to the DM for analysis. If the DM considers this set acceptable for final choice (in the sense of size), the decision-making process ends. Otherwise (the constructed Pareto set is "too wide"), one should try to acquire additional information in the form of a new quantum and then use it by analogy for further reduction of the set of selectable alternatives. In this case, there is a collection of information quanta for new vector criterion design and first one has to verify their consistency (see the details in Sect. 4.1). Note that generally such verification is reduced to the solution of an associated linear programming problem.

The sequential implementation of the above stages makes a cyclic process illustrated by Fig. 8.2. This process is repeated until the result satisfies the DM. The result is the current Pareto set whose size (as the DM believes) maximally fits the size of the set of selectable alternatives $C(X)$.

In some cases the DM is willing to lose certain quantities in terms of several criteria simultaneously in order to gain more in terms of a very important criterion. In other cases a loss in terms of a less important criterion cannot be compensated by an increase in terms of a single criterion but only in terms of several criteria simultaneously. In the general case, there may exist two groups of criteria with two disjoint index sets A and B such that the DM agrees to lose maximum w_j^* units in terms of the criteria f_j, for all $j \in B$, for gaining minimum w_i^* units in terms of the criteria f_i, for all $i \in A$. According to Definition 3.3, this means that *the group of criteria A is more important than the group of criteria B* with the two collections of positive parameters w_i^* and w_j^* for all $i \in A$ and all $j \in B$.

It is necessary to take into account an important aspect during quanta elicitation, as explained below. Theorem 3.1 states that the higher importance of the group of criteria A against the group of criteria B implies the higher importance of a wider group than A against a narrow group than B. Roughly speaking, a more important group can be always extended, whereas a less important group can be always reduced. And so, while eliciting information quanta one should make a more important group as narrow as possible and a less important group as wider as possible. Such information appears more essential, further facilitating higher reduction of the domain of compromise. In this sense, the best situation is where a certain criterion has higher importance against the group of all other criteria.

The new vector criterion is calculated on the basis of a general information quantum using Theorem 3.5. According to this theorem, from the initial collection of criteria $f_1, f_2, \ldots, f_m$ first we have to eliminate all less important criteria, i.e., the ones with indexes from the set B. And then the residual criteria are

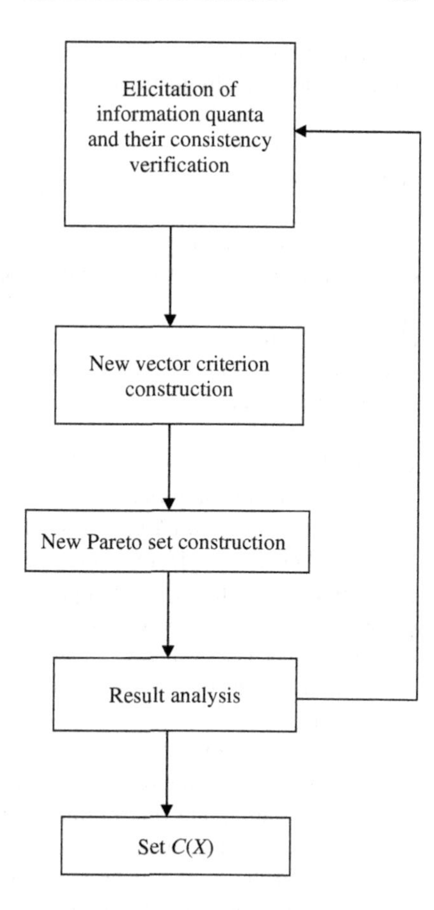

Fig. 8.2 The cyclic process

supplemented by the new criteria of the form $w_j f_i + w_i f_j$; their number is given by the product of the numbers of elements in the sets A and B.

Clearly, the total number of the resulting criteria can be much higher than in the original collection. For example, if the sets A and B consist of two and three elements, respectively, then we obtain 6 new criteria. Three less important criteria must be eliminated, but six new criteria are added. And the total number of the new criteria increases by 3.

If the group of more important criteria represents a singleton, the consideration of the existing information quantum does not enlarge the number of the resulting criteria (see Corollary 3.1), since the number of the new criteria coincides with the number of the old ones.

The theorems of Chaps. 3–5 have studied different collections of information quanta and stated the consistency conditions for them. If the available collection of information quanta does not satisfy any of these theorems, the new vector criterion can be calculated by the algorithms described in Sects. 5.3–5.4.

This method of Pareto set reduction based on a collection of information quanta has been theoretically justified in Chap. 6. In particular, Theorem 6.3 claims that in many cases where the set of feasible vectors is finite (this condition surely holds if the set of feasible alternatives is finite) one may construct the unknown set of nondominated vectors (ergo, the set of nondominated alternatives) with high precision using a finite collection of information quanta. Unfortunately, this result is not constructive in the sense that there exists no clear description which collection of information quanta to use. Moreover, the number of such quanta is also not specified. These issues strongly depend on the concrete form of the set of feasible alternatives and the criteria participating in the choice problem. Nevertheless, Theorem 6.3 is of crucial importance in theoretical terms, as it justifies the original axiomatic approach suggested in the book. As a matter of fact, it states that *one merely has to learn to elicit information quanta and to use them appropriately for solving the multicriteria choice problem. Using only such information it is possible to construct completely and precisely the set of nondominated alternatives for an arbitrary multicriteria choice problem from a rather wide class* with a finite set of feasible alternatives. If the latter set is infinite, then the set of nondominated alternatives or vectors can be approximated with a required precision using a finite collection of information quanta (see Theorem 6.2 for details).

Note that it is not always possible to get a clear answer from the DM during information quanta elicitation by questioning. "Is criterion i more important for you than criterion j?" The DM may find difficulty in replying this question and simply measure its confidence that it is true by a number (e.g., a share of 1). In this case, it is necessary to adopt the definition of a fuzzy information quantum about the DM's preference relation and the results of Chap. 7. But by supplying fuzzy information at the "input" of the suggested approach, one obtains a certain fuzzy set at the "output," and the final choice is performed within it. At the same time, usually in practice choice must be definite (crisp), and therefore a certain crisp subset has to be used. The extraction of this crisp set from the fuzzy set may cause difficulty. Unfortunately, there exist no universal recipes to form the crisp set from the fuzzy set, and each specific case requires detailed analysis. An example of such analysis and final choice recommendations can be found in Sect. 7.3.2.

8.3 Linear Scalarization Approach

8.3.1 Linear Combination of Criteria

The multicriteria problems can be solved using linear scalarization. According to this approach, it is necessary to assign somehow nonnegative (often positive) coefficients $\mu_1, \mu_2, \ldots, \mu_m$ such that $\mu_1 + \mu_2 + \cdots + \mu_m = 1$ (optional) and then to maximize the linear combination (weighed sum) of criteria $\sum_{i=1}^{m} \mu_i f_i(x)$

on the set X. In the sequel, the coefficients $\mu_1, \mu_2, \ldots, \mu_m$ will be called the *scalarization coefficients* or *weights*.

Such a solution method for the multicriteria problems appeared before the notion of Pareto optimality. It was actually pioneered in the 18th century by French mathematician and political scientist J.-C. de Borda for voting problems. In particular, the ranked preferential voting system suggested by him (the Borda count) determines the winner of an election by giving each candidate, for each ballot, a number of points corresponding to the number of candidates ranked lower. Once all votes have been counted the candidate with the most points is the winner. Let us consider it in detail.

The voting problem has a finite set of alternatives $X = \{x_1, \ldots, x_n\}$, referred to as candidates, and m voters who assign ranks for the elements of this set (i.e., place them in the descending order of their preferability). Ranking associates each candidate with a certain number starting from n (least preferable). In other words, on the set X it is necessary to define numerical functions $f_1, f_2, \ldots, f_m$, such that $f_k(x_j) = n_{kj}$, where n_{kj} means the rank of candidate x_j (starting from rank n) according to the opinion of voter k. By the *Borda count*, the winner is candidate i receiving the maximum sum $\sum_{k=1}^{m} f_k(x_i)$. This count employs the linear combination of the criteria with the unit weights.

Interestingly, outstanding Russian naval architect A. Krylov used a linear scalarization approach to assess the quality of products and services at the beginning of the 20th century.

Almost none of the researchers who solve applications-relevant problems using the linear scalarization approach consider whether their actions are valid or not. In terms of the multicriteria choice model, the linear combination of criteria is applicable under the following conditions:

(1) Axiom 1 (the exclusion axiom of dominated vectors) holds;
(2) There exist the scalarization coefficients $\mu_1, \mu_2, \ldots, \mu_m$ such that the preference relation $\succ$ is representable as a linear function, i.e.,

$$y \succ y' \quad \Leftrightarrow \quad \sum_{i=1}^{m} \mu_i y_i > \sum_{i=1}^{m} \mu_i y'_i \quad \text{for all} \quad y, y' \in R^m. \tag{8.1}$$

Indeed, by the exclusion axiom, condition (8.1) implies the following. Each vector for which there exists another vector with a greater value of the linear combination is eliminated from the set of selectable vectors, and therefore this set contains only the vectors maximizing the linear combination. Hence, under the two conditions above for any $C(Y)$ we have the inclusion

$$C(Y) \subset \left\{ y^0 \in Y | \sum_{i=1}^{m} \mu_i y_i^0 = \max_{y \in Y} \sum_{i=1}^{m} \mu_i y_i \right\}. \tag{8.2}$$

Note that the exclusion axiom is not universal in the sense that in some multicriteria choice problems it may fail.

The function $\sum_{i=1}^{m} \mu_i y_i$ satisfying (8.1) is called the *linear utility function*. This linear function is often said to form the binary relation $\succ$. Perhaps the existence of the linear utility function was first explored by the founders of game theory, J. von Neumann and O. Morgenstern. Their research was later continued by different authors. As it was established, the sufficient existence conditions of the linear utility function are rather restricting and hold for a relatively narrow class of preference relations (e.g., see [12]). They include the weak order requirement (which means irreflexivity and negative transitivity) of the relation $\succ$, as well as the so-called Archimedean axiom. As the binary relation $\succ$ defined by (8.1) is a cone relation, the necessary (yet, not sufficient) conditions for linear utility function existence are the compatibility axiom together with the transitivity and invariance of this relation with respect to a positive linear transformation. Another necessary condition for linear utility function existence consists in the transitivity of the indistinguishability relation $\approx$, which is defined by the equivalence

$$y \approx y' \Leftrightarrow \text{none of the relationships } y \succ y' \text{ and } y' \succ y \text{ holds.}$$

Analysis shows that this condition can be also characterized as rather "stringent."

Therefore, the linear utility function exists not in so many cases as it is believed by some researchers who use the linear scalarization approach for solving the multicriteria choice problems without the Edgeworth-Pareto principle.

Note another circumstance, too. Inclusion (8.2) means that the set of selectable vectors is contained in the set of the vectors maximizing the linear combination of criteria. If the latter represents a singleton, then this unique vector must be chosen for sure. However, simple examples show that this set may be rather wide (even coinciding with Y), and all its elements are nonequivalent in the similar sense as the non-equivalence of different Pareto optimal vectors. In this case, choice remains an open issue, and to find $C(Y)$ one has to involve additional information.

8.3.2 Linear Scalarization as a Choice Tool for Specific Pareto Optimal Vector

By accepting the Edgeworth-Pareto principle, it is possible to justify the applicability of the linear scalarization approach without requiring the existence of the linear utility function. Really, this principle dictates to perform choice within the Pareto set, and hence we may use the following Pareto optimality criterion in terms of the linear combination of criteria.

Theorem 8.1[1] *Let the set*

$$Y_* = \{y^* \in R^m | y^* \leqq y \text{ for some } y = (y_1, y_2, \ldots, y_m) \in Y\}$$

be convex.[2] *For a vector $y^0 \in Y$ to be Pareto optimal, a necessary condition is that there exists a collection of nonnegative numbers $\mu_1, \mu_2, \ldots, \mu_m, \sum_{i=1}^{m} \mu_i = 1$, such that*

$$\sum_{i=1}^{m} \mu_i y_i^0 = \max_{y \in Y} \sum_{i=1}^{m} \mu_i y_i. \tag{8.3}$$

Conversely, equality (8.3) *holding for some positive numbers $\mu_1, \mu_2, \ldots, \mu_m$ implies the Pareto optimality of the vector $y^0 \in Y$.*

Theorem 8.1 together with the Edgeworth-Pareto principle leads to the following result.

Theorem 8.2 *Let the exclusion and Pareto axioms hold. Assume that the set Y_* is convex. Then for any set of selectable vectors C(Y) we have inclusion* (8.2), *where the components of the vector $\mu = (\mu_1, \mu_2, \ldots, \mu_m)$ are nonnegative and their sum is 1.*

Hence, by varying the scalarization coefficients $\mu_1, \mu_2, \ldots, \mu_m$ within the above ranges and compiling all vectors that maximize the linear combination of criteria with these coefficients on the set *Y*, we construct a set that surely contains the desired set of selectable vectors. In the general case, this set may include the vectors resulting from the maximization of the linear combinations with several different collections of scalarization coefficients (i.e., with several vectors μ). Thus, if there is no guarantee that the linear utility function exists, theoretically some collection of linear combinations of criteria should be used to estimate the set *C*(*Y*).

Now consider the multicriteria choice problem where it is necessary to choose a single vector. According to inclusion (8.2), this vector can be obtained by assigning a single set of scalarization coefficients to the linear combination and maximizing this combination on the set *Y*.

As readers can see, there is some "discrepancy" between the necessary and sufficient conditions here: the former requires the nonnegativity of the scalarization coefficients μ_1, μ_2, …, μ_m, whereas the latter states that these coefficients are strictly positive.

Recall another fundamental notion of multicriteria optimization, namely, a proper efficient vector. A feasible point $x^* \in X$ is called *proper efficient* [14] with

[1]See, for example, [55].

[2]This assumption is true if the vector function *f* has componentwise concavity on the convex set *X*.

respect to a vector function f on a set X if it is efficient (Pareto optimal) and, besides, there exists a positive number A such that the inequality

$$\frac{f_i(x) - f_i(x^*)}{f_j(x^*) - f_j(x)} \leqq A$$

holds for all $i \in \{1, 2, \ldots, m\}, x \in X$, satisfying $f_i(x) > f_i(x^*)$, and for some $j \in \{1, 2, \ldots, m\}$ satisfying $f_j(x) < f_j(x^*)$. In this case, $f(x^*)$ is called *the proper efficient vector*.

If we confine analysis to the proper efficient vectors (instead of the Pareto optimal ones), then the above "discrepancy" can be partially neglected and the scalarization coefficients $\mu_1, \mu_2, \ldots, \mu_m$ can be assumed strictly positive. The more so because the "difference" between the sets of efficient and proper efficient vectors is not so considerable under the hypotheses of Theorem 8.1. (To be more precise, the latter set is dense in the former, see [55] for details).

However, the convexity of the set Y_* in Theorem 8.2 might not be ignored, since otherwise some Pareto optimal vectors are not obtained via the maximization of the linear combination with positive scalarization coefficients. An elementary example is the two-dimensional set $Y = \{y \in R^2 | y_1 \cdot y_2 \leqq 10,\ y_1 \geqq 1,\ y_2 \geqq 1\}$: all Pareto optimal points lie on the curved segment of its boundary, but the maximization of the linear combination with nonnegative scalarization coefficients yields only the limit points (1, 10) and (10, 1). The same reason explains the inapplicability of the linear scalarization approach if the set Y is finite and contains at least three vectors. Meanwhile, the linear combination of criteria underlies the well-known analytic hierarchy process (AHP), see [58]. This aspect should be remembered by those who use the AHP.

8.3.3 Normalization of Criteria and Assignment of Scalarization Coefficients

In real multicriteria choice problems the criteria are not abstract numerical functions, since they have specific meanings. To be more exact, the values of these functions express variables that belong to certain quantitative scales and are measured in certain units. As is well-known (see Sect. 2.4), the basic quantitative scales include the absolute scale, the ratio scale, the difference scale and the interval scale.

If the values of the criteria in the multicriteria choice problem are homogeneous (i.e., have the same scale and the same measurement units), then their linear combination surely makes sense. However, this situation is not widespread in applications, since the multicriteria problem can be often reduced to the single-criterion problem. For example, consider m different types of costs associated with the manufacturing process of some product. Here it seems unreasonable to solve the problem with m criteria, as the costs may be simply summed up to form a single criterion.

The multicriteria property of the choice problem appears due to the heterogeneity of the existing criteria: as a rule, their values belong to different scales and are measured in different units, which prevents from their combination using a single formula. A typical economic problem of this class is to obtain maximum profit in minimum time. Profit is measured using the ratio scale in RUB, USD, etc., whereas time using the difference scale in hours, days, years, etc. Is it possible to add the profit criterion to the time criterion? This sum becomes ill-posed. What is the way out of this situation?

A standard technique here (known as the normalization of criteria) consists reducing heterogeneous criteria to a unified scale. The resulting "artificial" scale has nothing in common with the "natural" quantitative scales mentioned above. This technique employs monotonic transformations that "equalize" the ranges of the criteria. The most widespread transformation of this class replaces an initial criterion f_i with the new criterion

$$\tilde{f}_i = \frac{f_i(x) - y_i^{\min}}{y_i^{\max} - y_i^{\min}},$$

where $y_i^{\max}$ and $y_i^{\min}$ are the maximum and minimum values, respectively, of the function f_i on the set X (of course, under the assumption that both exist). We will surely stay within the Pareto set after the maximization of the linear combination with positive scalarization coefficients and hence the normalized criterion should be rewritten as

$$\tilde{\tilde{f}}_i = \frac{f_i(x) - y_i^N}{y_i^{\max} - y_i^N},$$

where y_i^N means component i of the *nadir vector* defined by $y^N = (\min_{y \in P(Y)} y_1, \min_{y \in P(Y)} y_2, \ldots, \min_{y \in P(Y)} y_m)$.

The values of the transformed criteria have the range [0,1], which actually guarantees their "equalization." In addition, each of the transformations is positive linear (more specifically, affine) and hence strictly increasing. After such transformations the Pareto set remains the same, and therefore the normalization procedure can be performed together with the Edgeworth-Pareto principle and further usage of the linear combination of the transformed criteria.

The weak spot of the linear scalarization approach concerns the assignment of scalarization coefficients. It is often assumed that they characterize some "weight" or "importance" of a corresponding criterion. Still, none of the researchers has given a precise definition for these notions in the context of the linear combination, despite the existence of numerous methods to find them.

For example, the scalarization coefficients are chosen the same in the absence of any information about the priorities of criteria. If the criteria weights are strictly ordered, the scalarization coefficients are assigned so that they fill the interval [0, 1] uniformly. Another obtaining of the linear scalarization coefficients involves the pairwise comparison of the criteria weights and the analytic hierarchy process (AHP). In some cases the coefficients are assigned by experts. Each expert puts his own subjective meaning into the notion of criteria weights, and hence the linear scalarization approach is far from objectivity. Due to the above factors, this approach is treated as a heuristic approach to the multicriteria choice problems, i.e., rigorous substantiation is replaced with certain reasonable considerations. Another indication of its heuristic character is that one can hardly describe the class of problems where the linear scalarization approach surely yields the desired solution.

If the linear utility function exists for sure, then the scalarization coefficients must be chosen so that condition (8.1) holds. Clearly, this is unrealistic in practice. And so, the coefficients are often assigned using different tricks based on the uncertain notions of weight, importance, and so on.

Note that according to the linear scalarization approach a certain Pareto optimal vector is assigned. But the researcher does not know the exact relationship between the scalarization coefficients of the linear combination and different Pareto optimal vectors. Hence, regardless of the sophistication of scalarization coefficients assignment, there is no guarantee that the resulting Pareto optimal vector(s) would be really best in the given multicriteria choice problem.

8.3.4 Using Linear Scalarization at the Final Stage of Pareto Set Reduction

If we accept the Edgeworth-Pareto principle, then the solution of the original problem (i.e., the finding of the set of selectable vectors) can be interpreted as the Pareto set reduction to the set $C(Y)$. Such reduction is justified only if it uses some additional information about the DM's preference relation to "discard" the unsuitable Pareto optimal vectors. In this respect a convenient form of the additional information is a pair of Pareto optimal vectors in which the DM prefers one vector to the other. It is exactly the information in the form of quanta, see the previous chapters for its definition and usage in multicriteria choice problems.

Therefore, at the first stage of Pareto set reduction it is necessary to elicit consistent information about the DM's preference relation in the form of one or several quanta. Using this information and the earlier theorems or algorithms, we have to construct the new vector criterion $g = (g_1, g_2, \ldots, g_p)$, $p \geqq m$, satisfying the inclusions

$$C(X) \subset P_g(X) \subset P_f(X), \quad C(Y) \subset f(P_g(X)) \subset P(Y). \tag{8.4}$$

If the new Pareto set $P_g(X)$ is "too large" to perform final choice, then at the second stage we can apply the linear scalarization approach to implement its further reduction.

The next proposition justifies this combined approach.

Theorem 8.3 *Consider a certain finite consistent collection of information quanta about the DM's preference relation and the vector function* $g = (g_1, g_2, \ldots, g_p)$ *that participates in inclusions* (8.4). *Assume that the set* $X \subset R^n$ *is convex and the vector function* f *is componentwise concave on this set. Then for any set of selectable alternatives* $C(X)$ *we have the inclusion*

$$C(X) \subset \bigcup_{\mu} \left\{ x^* \in X \mid \sum_{i=1}^{p} \mu_i g_i(x^*) = \max_{x \in X} \sum_{i=1}^{p} \mu_i g_i(x) \right\}, \tag{8.5}$$

where the vector μ *satisfies the constraints*

$$\mu = (\mu_1, \mu_2, \ldots, \mu_p) \geq 0_p, \quad \sum_{i=1}^{p} \mu_i = 1. \tag{8.6}$$

☐ According to the axiomatic approach, the inclusion $C(X) \subset P_g(X)$ holds for any set of selectable alternatives $C(X)$.

Since the components of the initial vector function f are concave and the components of the new vector function g represent their N-combinations, the components of g are concave functions, too. It is easy to verify that in this case the set Y_* is convex. By Theorem 8.1, we obtain the inclusion

$$P_g(X) \subset \bigcup_{\mu} \left\{ x^* \in X \mid \sum_{i=1}^{p} \mu_i g_i(x^*) = \max_{x \in X} \sum_{i=1}^{p} \mu_i g_i(x) \right\},$$

where the vector μ has form (8.6). And the required result is established by integrating this inclusion with $C(X) \subset P_g(X)$. ■

Under the hypotheses of Theorem 8.3, let us also assume that the set of selectable alternatives (or vectors) is a singleton. Then the inclusion (8.5) can be replaced by the simple relationship

$$C(X) \subset \left\{ x^* \in X \mid \sum_{i=1}^{p} \mu_i g_i(x^*) = \max_{x \in X} \sum_{i=1}^{p} \mu_i g_i(x) \right\} \tag{8.7}$$

for some vector μ of form (8.6). Hence, in this case it suffices to choose a single vector of the scalarization coefficients for the linear combination of criteria and to solve the corresponding scalar maximization problem in order to construct the set that surely contains the desired "best" alternative.

The vector μ may have the same components, since the criteria priorities are taken into account during the formation and usage of the information quanta about the DM's preference relation. An additional correction of the priorities (if necessary) can be realized by redistributing the values of the scalarization coefficients for the linear combination of criteria. Here the following aspect should be kept in mind. The (originally nontrivial) assignment problem of the vector μ becomes even more complicated: the coefficients of this vector is associated with the components $g_1, g_2, \ldots, g_p$ of the new vector function that have no practical interpretation (in contrast to $f_1, f_2, \ldots, f_m$), since they represent the linear combinations of the initial criteria.

Moreover, in the general case some linear scalarization coefficients in (8.5) can be zero, since the vector μ has form (8.6). The existing zero coefficients make it necessary to consider the linear combinations of the initial criteria that may include from 1 to p terms. This feature introduces inconvenience during the implementation of the approach.

A refined version of Theorem 8.3 that does not suffer from this drawback can be obtained merely by imposing stronger conditions. In particular, the following result is true.

Theorem 8.4 *Consider a certain finite consistent collection of information quanta about the DM's preference relation that is taken into account using the vector function* $g = (g_1, g_2, \ldots, g_p)$. *Assume that the set* $X \subset R^n$ *is convex and compact, while the vector function* f *is concave and continuous on it, with at least one component of this function being strictly concave. Then for any set of selectable alternatives* $C(X)$ *we have*

$$C(X) \subset cl\left(\bigcup_{\mu}\left\{x^* \in X \mid \sum_{i=1}^{p} \mu_i g_i(x^*) = \max_{x \in X} \sum_{i=1}^{p} \mu_i g_i(x)\right\}\right), \tag{8.8}$$

where $cl(A)$ *denotes the closure of the set* A *and the vector* μ *satisfies the restrictions*

$$\mu = (\mu_1, \mu_2, \ldots, \mu_p) > 0_p, \quad \sum_{i=1}^{p} \mu_i = 1. \tag{8.9}$$

Theorem 8.4 follows directly from the results of the axiomatic approach and Corollary 5 from [55, p. 145].

In this theorem all components of the vector μ are positive, but the closure operation in the right-hand side of inclusion (8.8) slightly reduces the significance of this statement. If the selectable vector is unique and proper efficient, we may formulate the following proposition being convenient in practice.

Theorem 8.5 *Consider a certain finite consistent collection of information quanta about the DM's preference relation that is taken into account using the vector function* $g = (g_1, g_2, \ldots, g_p)$. *Assume that the set* $X \subset R^n$ *is convex, while the vector function* f *is concave on it. Then for any singleton* $C(X)$ *formed by the proper* *efficient alternative there exists a vector* μ *of form* (8.9) *such that inclusion* (8.7) *holds*.

Theorem 8.5 is based on the results of the axiomatic approach and the Geoffrion theorem [14] on the proper efficient point characterization using the linear combination of criteria in the case of concave criteria on a convex set of alternatives.

Therefore, under the above assumptions, at the second stage of the combined approach the unique "best" alternative can be sought for among the proper efficient points by solving a single maximization problem for the linear combination of the criteria $g = (g_1, g_2, \ldots, g_p)$ for some μ with positive components.

Each efficient point is proper efficient under definite constraints imposed on the vector criterion and the set of feasible alternatives. In this case, the properness requirement in Theorem 8.5 can be omitted. In particular, this holds in case with a linear vector function f and a polyhedral set X.

8.3.5 Combined Approach with Multiplicative Combination of Criteria

Along with the linear combination of criteria that allows reducing the initial multicriteria problem to the single-criterion one under the assumption that all criteria take positive values, researchers often employ the multiplicative combination $\prod_{i=1}^{m} f_i^{\mu_i}(x)$ where the vector μ has form (8.6) or (8.9). As a matter of fact, the maximization of the product of criteria implements the so-called fair compromise principle. Note that in theory of cooperative games the arbitration Nash solution is also constructed by maximizing the product of the utility functions of the players.

The results derived for the linear combination of criteria can be easily extended to the case of their multiplicative combination. To succeed, just use the chain of equivalences

$$\sum_{i=1}^{p} \mu_i \ln g_i(x^*) \geqq \sum_{i=1}^{p} \mu_i \ln g_i(x) \Leftrightarrow \ln \prod_{i=1}^{p} g_i^{\mu_i}(x^*) \geqq \ln \prod_{i=1}^{p} g_i^{\mu_i}(x)$$
$$\Leftrightarrow \prod_{i=1}^{p} g_i^{\mu_i}(x^*) \geqq \prod_{i=1}^{p} g_i^{\mu_i}(x)$$

and also the fact that the logarithm of a positive concave function is a concave function, too.

For instance, the next proposition follows directly from the aforesaid and Theorem 8.5.

Corollary 8.1 *Consider a certain finite consistent collection of information quanta about the DM's preference relation that is taken into account using the vector function* $g = (g_1, g_2, \ldots, g_p)$. *Assume that the set* $X \subset R^n$ *is convex, while the vector function* f *is concave and positive on it. Then for any singleton* $C(X)$ *formed by the* *proper efficient alternative there exist positive numbers* $\mu_1\mu_2,\ldots,\mu_p$ *such that the inclusion*

$$C(X) \subset \left\{x^* \in X | \prod_{i=1}^{p} g_i^{\mu_i}(x^*) = \max_{x \in X} \prod_{i=1}^{p} g_i^{\mu_i}(x)\right\}$$

takes place.

8.4 Combined Methods for Pareto Set Reduction

8.4.1 Using Uniform Metric

Recall that a point $x^* \in X$ is called *weakly efficient* (*Slater optimal*) with respect to f if there exists no $x \in X$ such that $f(x) > f(x^*)$. Each Pareto optimal (efficient) point is a weakly efficient one but not vise versa.

In 1966 Yu. Germeier [15] obtained the necessary and sufficient condition of weak efficiency. Later on, a series of authors rediscovered this result. According to the latter, a point $x^* \in X$ is weakly efficient if and only if there exist the vector $\mu = (\mu_1, \mu_2, \ldots, \mu_m) > 0_m, \sum_{i=1}^{m} \mu_i = 1$, such that on the set X the scalar function $\min_{i=1,2,\ldots,m} \mu_i f_i(x)$ reaches its maximum at the point x^*. The only prerequisite of the Germeier theorem is that all components of the vector criterion must have positive values at the given point, i.e., $f(x^*) > 0_m$. The function $\min_{i=1,2,\ldots,m} \mu_i f_i(x)$ is called the Germeier combination (in Russian literature) and the weighed Chebyshev metric (in English language literature), since in the case $\mu_1 = \mu_2 = \ldots = \mu_m$ it coincides up to a constant factor with the uniform metric suggested by P. Chebyshev.

Suppose that at the first stage of Pareto set reduction we have utilized one or several information quanta about the DM's preference relation, thereby constructing the new vector criterion $g = (g_1, g_2, \ldots, g_p)$, $p \geqq m$, that satisfies inclusions (8.4). The linear combination of positive functions with positive coefficients is a positive function. Each Pareto optimal point is weakly efficient. Hence, by combining the axiomatic approach with the Germeier theorem for the vector function g, we arrive at the following result.

Theorem 8.6 *Let the functions* $f_1, f_2, \ldots, f_m$ *have positive values on the set* X. *Consider a certain finite consistent collection of information quanta about the DM's preference relation that is taken into account using the vector function* $g = (g_1, g_2, \ldots, g_p)$. *Then for any set of selectable vectors* $C(Y)$ *we have the inclusion*

$$C(Y) \subset f(P_g(X)) \subset \bigcup_{\mu} \left\{ f(x^*) \in Y | \min_{i=1,2,\ldots,p} \mu_i g_i(x^*) = \max_{x \in X} \min_{i=1,2,\ldots,p} \mu_i g_i(x) \right\},$$

where the vector μ satisfies restrictions (8.9).

If the problem specifics dictate that the set $C(Y)$ must be a singleton, then using Theorem 8.6 we may obtain this set by solving the scalar (single-criterion) maximization problem for the function $\min_{i=1,2,\ldots,p} \mu_i g_i(x)$ on the set X under a fixed vector μ of form (8.9). Note that this maximization problem may have several solutions. In this case, one should involve additional considerations that would eliminate some candidates for the final choice. For example, apply the weighed sum of criteria, and so on.

The assignment problem for the weight coefficients $\mu_1, \mu_2, \ldots, \mu_p$ of the function $\min_{i=1,2,\ldots,p} \mu_i g_i(x)$ has been discussed in the previous section in the context of linear combination. And the conclusions fully apply to the Germeier combination. At the second stage it is reasonable to use the same coefficients $\mu_1, \mu_2, \ldots, \mu_p$, since the criteria priorities are taken into account during the formation and usage of the information quanta about the DM's preference relation. This approach totally removes the assignment problem. However, if there exists a vital need to correct the priorities of criteria, this can be implemented by adjusting the appropriate values of the coefficients.

Example 8.1 Consider the bicriteria choice problem with $Y = \{y^1, y^2, \ldots, y^5\}$, where $y^1 = (1, 2), y^2 = (2, 1.8), y^3 = (3, 1.7), y^4 = (4, 1.6), y^5 = (5, 1.3)$. Here all vectors have positive components and are Pareto optimal. Assume that by direct questioning we have established the higher importance of criterion f_1 against criterion f_2. Moreover, by losing 1 unit in terms of criterion f_2 the DM expects to gain at least 4 units in terms of criterion f_1. According to the axiomatic approach, this means the existence of an information quantum stating that criterion f_1 is more important than criterion f_2 with the parameters $w_1^* = 4, w_2^* = 1$. Using Theorem 2.5, construct the new vector criterion with the components $g_1 = y_1, g_2 = y_1 + 4y_2$. Find the image of the set of feasible alternatives in terms of the new vector criterion: $g(Y) = \{(1, 9), (2, 9.2), (3, 9.8), (4, 10.4), (5, 10.2)\}$. At the second stage, maximize the function $\min_{i=1,2} y_i$ on the this set to get the unique vector y^5, which is the "best" solution of the problem (the optimal choice).

In fact, there is another way to solve this problem. The inclusion $C(Y) \subset \hat{P}(Y) = f(P_g(X))$ holds by the Pareto principle, and hence before the second stage it is possible to eliminate from the set $g(Y)$ all vectors that are not Pareto optimal; and so, the function $\min_{i=1,2} y_i$ is then maximized on the truncated Pareto set $\{y^4, y^5\}$. Interestingly, by ignoring the available information quantum (skipping the first stage), we obtain the vector y^2 in the same uniform metric.

Just like for the Germeier theorem, the positivity of all criteria on their domain is a crucial requirement for Theorem 8.6, which might not be neglected. In the next theorem, this condition is replaced by the boundedness from above, which is not so restricting as positivity. This result takes place owing to the combination of the axiomatic approach with Corollary 3.1 from [17]. Note that the latter can be treated as the result of applying the Germeier theorem to the positive vector criterion $\hat{y} - g(x)$.

Theorem 8.7 *Consider a certain finite consistent collection of information quanta about the DM's preference relation that is taken into account using the vector function* $g = (g_1, g_2, \ldots, g_p)$. *Assume that the inequality* $g(x) < \hat{y}$ *holds for some* $\hat{y} \in R^p$ *and all* $x \in X$.*Then for any set of selectable vectors* $C(Y)$ *we have the inclusion*

$$C(Y) \subset \bigcup_{\mu} \left\{ f(x^*) \in Y | \max_{i=1,2,\ldots,p} \mu_i(\hat{y}_i - g_i(x^*)) = \min_{x \in X} \max_{i=1,2,\ldots,p} \mu_i(\hat{y}_i - g_i(x)) \right\},$$

where the vector μ *satisfies restrictions* (8.9).

The above remarks to Theorem 8.6 remain in force for Theorem 8.7, as well. Therefore, we do not repeat them here.

Usable result of this kind is based on the following new necessary and sufficient condition of weak efficiency point.

Lemma 8.1 *Fix an arbitrary number* α. *A point* $y^* \in Y = f(X)$ *is weakly efficient if and only if there exists a vector* $u \in R^m, \sum_{i=1}^m u_i = \alpha$, *such that*

$$\max_{i=1,2,\ldots,m} (u_i - y_i^*) \leqq \max_{i=1,2,\ldots,m} (u_i - y_i) \quad \text{for all} \quad y \in Y. \tag{8.10}$$

□ Necessity. Let y^* be an arbitrary weakly efficient vector. Introduce the vector u with the components

$$u_i = y_i^* - \frac{1}{m}\left(\sum_{j=1}^m y_j^* - \alpha\right), \quad i = 1, 2, \ldots, m; \sum_{i=1}^m u_i = \alpha.$$

Conjecture the opposite, i.e., there exists a point $y \in Y$ satisfying $\max\limits_{i=1,2,\ldots,m} (u_i - y_i^*) > \max\limits_{i=1,2,\ldots,m} (u_i - y_i)$. Hence, for any i we have the inequality

$$u_i - y_i < \max_{i=1,2,\ldots,m} (u_i - y_i^*) = -\frac{1}{m}\sum_{j=1}^m y_j^* + \frac{\alpha}{m}.$$

Substitute the corresponding values of the components u_i into the left-hand side of this inequality to get $y_i^* < y_i, i = 1, 2, \ldots, m$. This obviously contradicts the weak efficiency of y^*.

Sufficiency. Take the vector u that satisfies the hypotheses of Lemma 8.1 and inequality (8.10). Again, we will prove by contradiction. If the vector y^* is not weakly efficient, then there exists $y \in Y$ such that $y^* < y$. In this case, we have $u_i - y_i^* > u_i - y_i, i = 1, 2, \ldots, m,$ which directly gives $\max_{i=1,2,\ldots,m}(u_i - y_i^*) > \max_{i=1,2,\ldots,m}(u_i - y_i)$. The last inequality contradicts (8.10). ∎

The combination of the axiomatic approach with Lemma 8.1 yields the following result.

Theorem 8.8 *Consider a certain finite consistent collection of information quanta that is taken into account using the vector function* $g = (g_1, g_2, \ldots, g_p)$. *Fix an arbitrary number* α. *Then for any set of selectable vectors* $C(Y)$ *we have the inclusion*

$$C(Y) \subset \bigcup_u \left\{ f(x^*) \in Y \,\middle|\, \max_{i=1,2,\ldots,p}(u_i - g_i(x^*)) = \min_{x \in X} \max_{i=1,2,\ldots,p}(u_i - g_i(x)) \right\},$$

where the vector $u = (u_1, u_2, \ldots, u_p)$ *such that* $\sum_{i=1}^p u_i = \alpha$.

If the set $C(Y)$ represents a singleton, then by Theorem 8.8 it is contained among the solutions of the minimization problem for the function $\max_{i=1,2,\ldots,p}(u_i - g_i(x))$ on the set X. We may specify numbers $u_1, u_2, \ldots, u_p$ by the desired ("ideal") values for each criterion, e.g., their maximum values on the set X, i.e., $u_i = \sup_{x \in X} g_i(x), i = 1, 2, \ldots, p$. By minimizing the maximum function $\max_{i=1,2,\ldots,p}(u_i - g_i(x))$, we obtain the weakly efficient vector from the set $g(X)$ that is uniformly closest to the vector u. This idea to choose the feasible vector having the shortest distance to some "ideal" vector (or even to some set of "ideal" vectors) underlies *goal programming*. In the forthcoming subsection we employ the same idea of approximation but on the basis of the Euclidean metric.

Example 8.2 Consider the bicriteria choice problem from Example 8.1 with the same information quantum. The difference is that at the second stage we will perform scalarization according to Theorem 8.8. To this end, choose $u = (5, 10.4)$ as the "ideal" vector (actually, it contains the maxima of the components of the vectors $\{y^4, y^5\}$ from the new Pareto set). As easily seen, in this case we again obtain vector y^5 as the "best".

Note that in Theorem 8.8 the parameter α can be chosen arbitrarily. If the vector u is defined as the nadir vector with the components $u_i = \inf_{x \in P_g(X)} g_i(x)$, $i = 1, 2, \ldots, p$, then the minimization of the function $\max_{i=1,2,\ldots,p}(u_i - g_i(x))$ on the set X yields the weakly efficient vector having the uniformly largest distance to the nadir vector.

8.4.2 Using Euclidean Metric

Assume that on the set X all criteria are bounded from above and introduce the set

$$U = \left\{ u \in R^m | u_i > \sup_{x \in X} f_i(x), i = 1, 2, \ldots, m \right\},$$

which is often called the set of *ideal* or *utopian vectors.*

Let us solve the multicriteria choice problem using goal programming based on the idea that the "best" vector of the set Y is the one closest to the set U. The distance between vectors $a, b \in R^m$ we will measure by the standard Euclidean metric $\|a - b\| = \sqrt{\sum_{i=1}^{m} (a_i - b_i)^2}$.

Lemma 8.2 *Assume that the set $X \subset R^n$ is convex and the numerical functions $f_1, f_2, \ldots, f_m$ are bounded from above and componentwise concave on it. A point $x^* \in X$ is proper efficient with respect to f if and only if there exists a vector $u \in U$ such that*

$$\|u - f(x^*)\| = \min_{x \in X} \|u - f(x)\|.$$

□ First, observe that under the hypotheses of this lemma the set

$$Y_* = \{ y^* \in R^m | y^* \leqq y \text{ for some } y \in Y \}$$

is convex; moreover, the sets of efficient (Pareto optimal) vectors (as well as the sets of proper efficient vectors) on the sets Y and Y_* coincide (see, for example, [55, p. 99]).

Necessity. Let the vector $y^* = f(x^*) \in Y$ be proper efficient. In this case, by the Geoffrion theorem [14] there exists a vector μ with components $\mu_i > 0_m, i = 1, \ldots, m, \sum_{i=1}^{m} \mu_i = 1$ such that on the set Y_* the linear function $\sum_{i=1}^{m} \mu_i y_i$ reaches its maximum at the point $y^* \in Y_*$. In other words, the vector μ is a normal vector of the supporting hyperplane for the convex set Y_* at this point. Since the components of the vector μ are positive, it is possible to find a positive number α such that $y^* + \alpha \cdot \mu \in U$. Choose $u = y^* + \alpha \cdot \mu$ as the vector $u \in U$ satisfying the hypotheses of Lemma 8.2. Then the above hyperplane is also supporting for the closed ball with radius $\|u - y^*\|$ and the center located at the point u. This means that the point y^* is the closest point of the set Y_* (ergo, of the set Y) to the point u.

Sufficiency. Denote by $y^* = f(x^*) \in Y$ the point that implements the minimum distance from the set Y to some vector $u \in U$. Clearly, among all points of the set Y_* the point y^* has the minimum distance to u. Let L be the hyperplane that passes through the point y^* and is supporting for the ball with radius $\|u - y^*\|$ and the center located at the point u. The relative interiors of the convex set Y_* and this ball do not intersect. Therefore, by Theorem 11.3 [57] the set Y_* and the ball can be separated by a certain hyperplane passing through their common point y^*. The ball has the unique supporting hyperplane L at the point y^*, and so this hyperplane is supporting for the set Y_*, too. Consequently, the linear function $\sum_{i=1}^{m}(u_i - y_i^*)y_i$ with the positive coefficients reaches its maximum on the set Y_* (ergo, on the set Y) at the point y^*. This implies the proper efficiency of the point y^*. ∎

Now, analyze the combined approach that consists of two sequential stages, namely,

(1) the Pareto set reduction using one or several consistent information quanta,
(2) the minimization of the distance between the set $g(X)$ and a preselected point of the "ideal" set $U' = \left\{u \in R^p | u_i > \sup_{x \in X} g_i(x), i = 1, 2, \ldots, p\right\}$.

Theorem 8.9 *Consider a certain finite consistent collection of information quanta that is taken into account using the vector function* $g = (g_1, g_2, \ldots, g_p)$. *Assume that the set* $X \subset R^n$ *is convex, while the functions* $f_1, f_2, \ldots, f_m$ *are bounded from above and componentwise concave on it. Then for any set of selectable vectors* $C(Y)$ *we have the inclusion*

$$C(Y) \subset cl\left(\bigcup_{u \in U'} \left\{f(x^*) \in Y | \|u - g(x^*)\| = \min_{x \in X} \|u - g(x)\|\right\}\right), \tag{8.11}$$

where $cl(A)$ *denotes the closure of the set* A.

Theorem 8.9 combines the axiomatic approach with Lemma 8.2 and the fact that, under the hypotheses of this theorem, the set of proper efficient vectors is dense in the set of efficient vectors (see [55, p. 140] for details). In other words, the closure of the set of proper efficient vectors contains the set of efficient vectors.

As we see, Theorem 8.9 has the restricting assumptions that may be crucial for its force. In particular, if the new Pareto set is finite (which surely holds under a finite set Y), then the scalarization approach based on the Euclidean metric is generally not correct. This circumstance should be kept in mind by those who solve the applications-relevant multicriteria choice problems using the idea that the "best" vector of the set Y is the one closest to a certain vector from the set U. An illustrating example below explains it.

Example 8.3 Consider $Y = \{y^1, y^2, y^3\}$, where $y^1 = (1, 3), y^2 = (1.5, 1.5)$, $y^3 = (3, 1)$. Clearly, all vectors are Pareto optimal; however, the second vector is not the closest one (in terms of the Euclidean metric) from the points y^1, y^2, y^3 to the set $\{u \in R^2 | u_1, u_2 > 3\}$. Note that the vector y^2 is closest, e.g., to the vector $(3.1, 3.1)$ in terms of the uniform metric.

8.5 Customs Duty Optimization

8.5.1 Problem Statement

Suppose that Russia trades a certain product with other countries on its domestic market. The product is manufactured inland and also abroad. This trade yields the national budget income

$$S = t_d xp + \tau qy + t_m(1+\tau)qy, \tag{8.12}$$

where

- t_d is the value-added tax (VAT) for the product manufactured in Russia (presently, $t_d = 18\%$);
- t_m denotes VAT for the imported products and services (presently, this tax has three rates for three different categories of products and services, namely, 0% for the high importance (crucial) products, e.g., the components and devices for space industry; 10% for the medium importance products, e.g., food items and baby goods; and 18% for the low importance (common) products and services);
- x gives the domestic output of the product;
- p specifies the domestic price of the product;
- τ is the import duty of the product (in the sequel, we will analyze the ad valorem duties only; note that the so-called specific duties are also considered in the literature, e.g., in [22]);
- q means the foreign price of the product (under the assumption that the importer has 0% VAT in the country of product purchase; in other words, the quantity q is much smaller than the retail price in a corresponding country);
- y indicates the import volume

Without loss of generality, let $t_d = t_m = t$, since for most products and services VAT is defined in the same way for the inland products and for the imported products.

The importer of the product makes the profit

$$D = y[p - (1+\tau)(1+t)q]. \tag{8.13}$$

According to the Russian Tax Code (see [63, Chapter 21, articles 150, 153, 164 and 166], the products and services imported in Russia (with some exceptions) are liable to VAT; the taxable base is the customs value including the import duty. Different countries introduce different VAT rates, and in some countries (e.g., USA, Japan) VAT does not exist at all. Most European countries have high rates (e.g., 25% VAT in Norway). China, India, and Arab countries use 10% VAT. For the Russian budget, this tax yields hundreds of billions RUR.

A consuming country faces the problem of import regulation (actually, a choice problem) depending on the current market situation by establishing an appropriate import duty or quota system (in this book, we will not consider quota systems, as optimal duties and optimal quotas are interconnected, both representing governmental regulation tools). The objective of such regulation can be the tax revenues or customs charges (8.12) or the share of domestic manufacturers in the market sales. The importer's objective is profit (8.13) or the number of jobs in the countries that manufacture the imported product (which is assumed directly proportional to the import volume). To control its objective, the importer varies the import volume appropriately.

In the sequel, by assumption the global supply price q is constant and not affected by the domestic consumption of a country. Moreover, we will not analyze substitute products or accompaniments. These features essentially simplify the problem and its solution. The market of the product is competitive, which allows adopting the Walras equilibrium concept to describe the domestic price of the product. Note that the cases of the price increase due to the monopolies or oligopolies on the market are also not considered.

Let us pose the three problems as follows.

Problem A: find the Pareto set for the two objective functions S and D under $y \geqq 0, \tau \geqq 0, D \geqq 0$.

The domestic output x in this problem is assumed constant. The solution of Problem A defines the capabilities of the mutual strategies of the consuming country and importer. The Pareto set gives the domain of compromise in trade policy, and the choice within this set allows to establish the mutually beneficial strategy (at the negotiating table or, e.g., by the trial-and-error method).

Problem B: find the Pareto set for the three objective functions S, D, and $Y = y$ under $y \geqq 0, \tau \geqq 0, D \geqq 0$.

The domestic output x in this problem is assumed constant, too. The solution of Problem B defines the capabilities of the mutual strategies of the consuming country, importer companies and exporting countries: here the additional objective of optimization is the increasing (or nondecreasing) number of jobs. Such a statement becomes topical, e.g., in the following situation. The import volume is so large that the trade unions of the consuming country (or the export control authorities of the exporting country) require appropriate countermeasures from the government for import duty establishment. For instance, the constraints imposed by the Russian government on the US chicken imports (the so-called Bush's legs) caused series problems at the poultry farms in Maryland, which initiated the political response of the US government. At the same time, it seems that the Pareto set construction would yield an acceptable compromise solution for both sides.

Problem C: find the Pareto set for the three objective functions S, D, and $X = x$ under $y \geqq 0, \tau \geqq 0, D \geqq 0$.

The domestic output x in this problem is assumed variable, which corresponds to the supply function of the domestic manufacturers. This statement analyzes the protectionist behavior capabilities in order to improve the competitiveness of the domestic manufacturers or simply to increase the domestic output (i.e., the government of the consuming country is concerned about the number of jobs in it).

Let us accept the notation $\Omega = \{(y, \tau) | y \geqq 0, \tau \geqq 0, D \geqq 0\}$.

In Problems A-C the domestic price of the product is constant and hence we may choose it as the money unit: $q = 1$. Similarly, since the statements of Problems A and B involve a constant domestic output, we choose $x = 1$ as the measurement unit for the import volume and consumption volume. In addition, suppose that an internal customer spends a fixed sum M from its budget on this product. Then the demand curve is defined by

$$p(x+y) = M. \tag{8.14}$$

Under the above assumptions, for Problems A and B we have the three objective functions

$$S = \frac{txM}{x+y} + [\tau + t(1+\tau)]qy, \tag{8.15}$$

$$D = y\left[\frac{M}{x+y} - (1+\tau)(1+t)q\right], \tag{8.16}$$

$$Y = y. \tag{8.17}$$

Actually, the objective function (8.17) does not participate in Problem A. The variables are $y \geqq 0, \tau \geqq 0$, while M acts as a positive parameter. We will choose $M = 6$ for the calculations and figures below.

8.5.2 *Solution of Posed Problems*

Find the Pareto set in *Problem A*. According to (8.16), the import profitability condition $D \geqq 0$ is expressed by the inequality $\tau \leqq \frac{M}{q(1+t)(x+y)} - 1$, see the domain Ω in Fig. 8.3 for $t = 0.18$. Here the vertical segment gives the solution of Problem A. Since this problem involves only two objective functions each having two variables, it is possible to apply the well-known approach based on the Edgeworth box.

Lemma 8.1 *The set of locally Pareto optimal alternatives that compromise between the consuming country and the importer is contained in the set of points* $\left\{(\bar{y},\tau)|0 \leqq \tau \leqq \frac{1}{1+t}\sqrt{\frac{M}{qx(1-t)}} - 1\right\}$, *where* $\bar{y} = \sqrt{\frac{(1-t)Mx}{q}} - x$.

□ A point (y, τ) lies beyond the set of locally Pareto optimal points if at this point the contour curves of functions (8.15) and (8.16) intersect. Otherwise, when the gradients of these functions are collinear, the set of such points contains the locally Pareto optimal set. The above collinearity condition has the explicit form

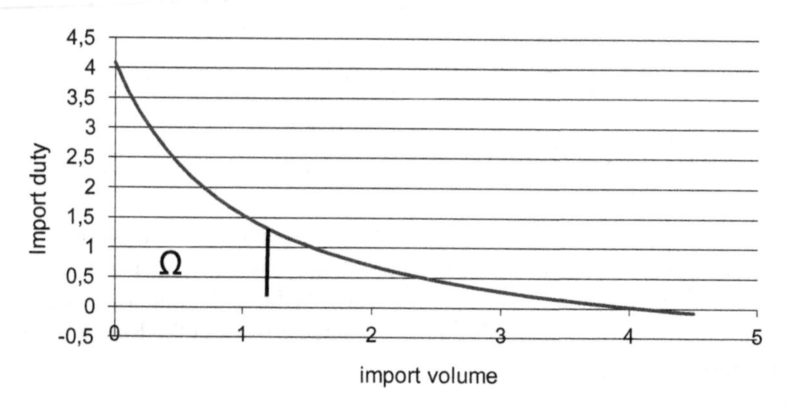

Fig. 8.3 Set Ω

$$-\frac{txM}{(x+y)^2}+[\tau+t(1+\tau)]q=\vartheta\left[\frac{M}{x+y}-(1+\tau)(1+t)q-\frac{yM}{(x+y)^2}\right],$$
$$(1+t)qy=-\vartheta qy(1+t).$$

The second equality implies $\vartheta=-1$. Then the first equality yields the quadratic equation

$$q(x+y)^2-M(x+y)+m(y+tx)=0,$$

whose positive solution is the optimal import volume

$$\bar{y}=\sqrt{\frac{(1-t)Mx}{q}}-x.$$

Recall that we are interested only with the points of the set Ω. Hence, the import duty must be varied within the range

$$0\leqq\tau\leqq\frac{1}{1+t}\sqrt{\frac{M}{qx(1-t)}}-1.$$

∎

Computer simulations show that the set mentioned in Lemma 8.1 is the Pareto set.

Remark 8.1 The optimal import volume $\bar{y}$ is independent of the tax duty. And so, the substitution of $\bar{y}$ into (8.15) and (8.16) gives the following linear dependencies on τ:

$$S(\bar{y}) = t\sqrt{\frac{qxM}{1-t}} + [(1+t)\tau + t]\sqrt{qxM(1-t)} - qx,$$

$$D(\bar{y}) = \left[\sqrt{qxM(1-t)} - qx\right]\left[\sqrt{\frac{M}{(1-t)qx}} - (1+t)(1+\tau)\right].$$

The first dependence is increasing, while the second decreasing (both have the same slope ratio). In other words, the import duty increase by 1% enhances the consuming country's income and simultaneously reduces the importer's income by the same quantity (which depends on many characteristics of the Russian economy and the global prices).

Remark 8.2 According to the approach from [56] (the model with $t_m = 0\%$), the importer adopts the strategy $y^* = \arg\max D(y, \tau)$ for a fixed import duty τ, and then the consuming country defines the optimal duty τ^* by the import volume y^*. Using the current notation, we have

$$\frac{\partial S}{\partial y} = \frac{M}{1+y} - 1 - \tau - \frac{My}{(1+y)^2} = 0.$$

Hence, $y^* = \sqrt{\frac{M}{1+\tau^*}} - 1$ and $\tau^* = \frac{M}{(1+y^*)^2} - 1$. The consuming country's profit is achieved on the curve

$$S^*(\tau) = 1.18\sqrt{M(1+\tau)} + \tau\left(\sqrt{\frac{M}{1+\tau}} - 1\right).$$

The highest profit corresponds to $\tau^* \approx 2.29$ (for $M = 6$). In this case, we obtain $y^* \approx 0.35$. The substitution of these values into the objective functions S and D yields $S^* \approx 1.6$, $D^* \approx 0.4$. On the other hand, the Pareto set allows choosing a compromise alternative for $\hat{\tau}$ from the interval [0, 1.7]. According to the approximate estimates, S grows from 0.5 to 2.5 while D reduces from 2.1 to 0 as we increase $\hat{\tau}$. In particular, for $\hat{\tau} = 0.9$ we have $S = 1.6$ and $D = 1$. Thus, the trade policy based on the Pareto set is much better.

The solution of *Problem B*. The third objective function $Y = y$ reflecting the interests of the exporting countries slightly modifies the construction of the Pareto set within $\Omega = \{(y, \tau)|y \geqq 0, \tau \geqq 0, D \geqq 0\}$.

Using the definition of a Pareto optimal alternative and the direct calculations, we establish that in Problem B the Pareto set acquires the form

$$\left\{(\bar{y}, \tau)|y \geqq \sqrt{\frac{(1-t)xM}{q}} - x, \tau \geqq 0, D \geqq 0\right\}.$$

Proceed to Problem C. As before, the main objective functions have expressions (8.12) and (8.13), but equality (8.14) now incorporates the variable x (the domestic output of the product) determined by the supply function. This function will be assumed linear (note that in practice one should employ statistical identification methods in order to obtain proper economic interpretations). And so, let the supply function be

$$p = ax + b,$$

where $a > 0,\ b > 0$. The substitution of $x = (p - b)/a$ into (8.14) gives the domestic price

$$p = \frac{b - ay}{2} + \sqrt{\frac{(b - ay)^2}{4} + aM}.$$

Here we choose the positive root of the associated quadratic equation due to the constraint $p > 0$.

Again, reduce the number of the parameters by introducing special measurement units for the money and sales volume: $q = 1$ and $a = 1$. Then the domestic price and the domestic output are defined by the formulas

$$p = \frac{b - y}{2} + \sqrt{\frac{(b - y)^2}{4} + M}, \quad x = -\frac{b + y}{2} + \sqrt{\frac{(b - y)^2}{4} + M}.$$

Their substitution into the objective functions (8.12)–(8.13) yields

$$S = txp + \tau y + t(1+\tau)y = \tau y + t(1+\tau)y + t\left[M - \frac{y}{2}\left(b - y + \sqrt{(b - y)^2 + 4M}\right)\right],$$

$$D = \frac{y}{2}\left[b - y + \sqrt{(b - y)^2 + 4M} - 2(1+\tau)(1+t)\right].$$

Now, calculate the corresponding derivatives:

$$\frac{\partial S}{\partial y} = (t + t\tau + \tau) + \frac{t}{2}\left[2y - b - \sqrt{(b - y)^2 + 4M} - \frac{y(y - b)}{\sqrt{(b - y)^2 + 4M}}\right], \frac{\partial S}{\partial \tau} = y(t + 1),$$

$$\frac{\partial D}{\partial y} = \frac{1}{2}\left[b - 2y + \sqrt{(b - y)^2 + 4M} + \frac{y(y - b)}{\sqrt{(b - y)^2 + 4M}} - 2(1+\tau)(1+t)\right], \frac{\partial D}{\partial \tau} = -y(t + 1),$$

$$\frac{\partial X}{\partial y} = \frac{1}{2}\left[\frac{y - b}{\sqrt{(b - y)^2 + 4M}} - 1\right], \frac{\partial X}{\partial y} = 0.$$

Further analysis requires some information about the parameter b of the domestic supply function. If b is smaller than 1, the product can be manufactured domestically in sufficient quantities without serious problems. Since the global unit price is 1, then for $b<1$ the domestic production is possible only using state-of-the-art technologies without the shortage of production factors. In the case $b>4$ it seems even unreasonable to boost the production. Therefore, we believe that $1<b<4$.

Since $\frac{\partial S}{\partial \tau} = y(t+1)$ and $\frac{\partial D}{\partial \tau} = -y(t+1)$, let us first estimate the Pareto set for the objective functions S and D. Their gradients must be collinear at all points of this set, hence

$$2y - b - \sqrt{(b-y)^2 + 4M} - \frac{y(y-b)}{\sqrt{(b-y)^2 + 4M}} + \frac{2}{1-t} = 0.$$

Here b, M, and t indicate some parameters. For $M = 6$ and $t = 0.18$, computer simulations (e.g., in MS EXCEL) show that the above equation has the unique positive solution with the almost linear dependence on the parameter b:

$$\breve{y} \approx 0.3355b + 1.633.$$

On the plane (y, τ) the condition $D > 0$ is equivalent to the inequality

$$\tau < \frac{1}{2(1+t)}\left[b - y + \sqrt{(b-y)^2 + 4M}\right] - 1 = \breve{\tau}.$$

Due to $\frac{\partial X}{\partial \tau} < 0$, the consideration of the third objective function X leads to the Pareto set in the form of the curvilinear trapezoid $0 \leqq y \leqq \breve{y}, 0 \leqq \tau \leqq \breve{\tau}$ depending on the parameters b, M, and t.

8.5.3 *Pareto Set Reduction*

Get back to Problem A. As follows from (8.12) and (8.13), the objective functions S and D are linear for fixed y and their slope ratios represent the opposite numbers. Therefore, if we fix $y = \bar{y}$ using Lemma 8.1 to construct the segment containing the Pareto set, then the resulting linear functions of the variable τ have the slope rations $\bar{y}$ and $-\bar{y}$.

Suppose that by negotiations (or due to the current situation) the consuming country puts the necessary leverage on the importer, compelling it to lose maximum $w_D > 0$ units in terms of the criterion D, and by turn expects to gain minimum $w_S > 0$ units in terms of the criterion S. This information specifies a definite quantum. According to Theorem 2.5, we arrive at the new bicriteria problem with the objective functions S and $w_D S + w_S D$. If $w_D = w_S$, the slope ratio of the second objective linear function is 0, i.e., this function is a constant. The Pareto set with respect to the criteria S and the constant function coincides with the set of the maximum points of the function S on the interval specified by Lemma 8.1, i.e., with the point

$$\bar{y} = \sqrt{\frac{(1-t)xM}{q}} - x, \bar{\tau} = \frac{1}{1+t}\sqrt{\frac{M}{qx(1-t)}} - 1.$$

The same result is obtained in the case $w_D > w_S$, as the slope ratio of the linear function $w_D S + w_S D$ under variable τ has the same sign as that of the function S. If the inequality $w_D < w_S$ holds, then the new Pareto set coincides with the initial counterpart. In other words, then the existing information quantum does allow to reduce the Pareto set.

By analogy we may consider the symmetrical situation where the importer puts leverage on the consuming country. As a result, the latter is willing to lose maximum w_s units in terms of the criterion S, also having no objections that the importer gains minimum w_D units in terms of the criterion D. In this case, we obtain the new bicriteria problem with the objective functions $w_D S + w_S D$ and D. If $w_D \leqq w_S$, then the Pareto set is reduced to the point $\bar{y} = \sqrt{\frac{(1-t)xM}{q}} - x, \bar{\tau} = 0$; otherwise, no reduction takes place.

Problem B involves the three criteria S, D and $Y = y$. In theory, according to the existing information, a certain pair of the criteria may have the property that a loss in terms of one criterion gives a gain in terms of the other. Moreover, it is quite possible that a loss (or a gain) relates to a pair of criteria while a gain (a loss, respectively) to the third criterion. Also a realistic scenario is where the consuming country compels the importer to lose in terms of the criterion D, also expecting some gains in terms of the criteria S and Y. Using information about the criteria for potential losses and gains and their amounts, it is possible to reduce the Pareto set based on Theorem 2.5.

For example, assume that the consuming country and the importer are willing to lose maximum 1 unit in terms of the criterion Y for gaining minimum 0.3 units in terms of the criteria S and D, i.e., $w_S = w_D = 0.3, w_Y = 1$. As the calculations show, the corresponding Pareto set has the form

$$\left\{(y,\tau)|\sqrt{\frac{(1-t)xM}{q}} - x \leqq y \leqq \sqrt{\frac{(1-t)xM}{q-0.6}} - x, \tau \geqq 0, D \geqq 0\right\}.$$

Consider Problem C. Recall that the associated Pareto set is the curvilinear trapezoid $\left\{(y,\tau)|0 \leqq y \leqq \breve{y}, 0 \leqq \tau \leqq \breve{\tau}\right\}$, where $\breve{y} \approx 0.3355b + 1.633$ and

$$\breve{\tau} = \frac{1}{2(1+y)}\left[b - y + \sqrt{(b-y)^2 + 4M}\right] - 1.$$

Suppose that the consuming country is able to compel the importer to lose $w_D > 0$ units in terms of the criterion D (the importer's profit) in order to increase the value of the criterion S (the tax revenues of the national budget) by maximum $w_S > 0$ units. To construct the new Pareto set using Theorem 2.5, we have to solve the tricriteria problem with the objective functions S, $w_D S + w_S D$, and x (by assumption, the domestic production development, in particular, the number of jobs, has lower priority than the tax revenues).

Similarly, in the opposite situation where the consuming country is willing to lose w_S units in terms of the criterion S while the importer gains w_D units in terms of the criterion D, we get the tricriteria problem with the objective functions $w_D S + w_S D$, D and x.

In another realistic scenario the consuming country is willing to lose w_x units in terms of the domestic production in order to gain w_S units in terms of the criterion S. In this case, the criteria are the three functions S, D, and $w_x S + w_S x$. If the consuming country is concerned with the number of jobs and the domestic production development, then the criteria are the functions x, D, and $w_x S + w_S x$.

Furthermore, in Problem C (like in Problem B) the losses (or gains) may affect two criteria of the three existing ones S, D, and x.

Consider the following illustrative example with numerical data, which describes one of the last scenarios. Let the national budget income S have the highest importance. For $M = 6$, $t = 0.18$, $b = 3$, $w_x = 1$, and $w_S = 2$, the functions S, D and $G = w_x S + w_S x$ acquire the form

$$S = (0.18 + 1.18\tau)y + 0.18\left[6 - \frac{y}{2}\left(3 - y + \sqrt{(3-y)^2 + 24}\right)\right],$$

$$D = \frac{y}{2}\left[3 - y + \sqrt{(3-y)^2 + 24} - 2.36(1+\tau)\right],$$

$$G = (0.18 + 1.18\tau)y + 0.18\left[6 - \frac{y}{2}\left(3 - y + \sqrt{(3-y)^2 + 24}\right)\right] + y\left[3 - y + \sqrt{(3-y)^2 + 24}\right].$$

To construct the Pareto set, we calculate the gradients of the three functions and verify when the origin belongs to the triangle formed by their ends. As a result, we find the set

$$\left\{y + [1.69, 2.64], \tau \in \left[0, \frac{1}{2.36}\left(3 - y + \sqrt{(3-y)^2 + 24}\right)\right]\right\}.$$

Obviously, this set is appreciably narrower than the set obtained in the initial statement of Problem C without consideration of the information quanta.

8.6 Production Output Increase Problem with Resource Costs

8.6.1 Problem Statement

Assume that a certain firm seeks to increase its production output, reducing simultaneously the resource costs. Clearly, a higher output cannot be achieved without consuming additional resources, and so output maximization contradicts resource reduction. Such a situation is typical for multicriteria optimization. Therefore, it seems reasonable to formalize this problem as some multicriteria choice problem. Let us pass to this formalization.

In many economic problems the relationship between the production output and resource consumption is modeled by the power production functions $z = ax_1^{\alpha_1} \cdot \cdots \cdot x_n^{\alpha_n}$. Here z denotes the output, $x_1, \ldots, x_n$ are the volumes of consumed resources, and $a, \alpha_1, \ldots, \alpha_n$ represent some positive parameters.

Our analysis will be confined to two types of resources, namely, the labor resources and the basic production assets. Thus, the production output z is defined by the labor costs x_1 and the capital x_2 according to the formula $z = ax_1^{\alpha_1}x_2^{\alpha_2}$ for some positive parameters $\alpha, \alpha_1, \alpha_2, \alpha_1 + \alpha_2 < 1$.

Formulate the following multicriteria problem. The set of feasible alternatives consists of all pairs $\{(x_1, x_2)\}$ such that $x_1, x_2 > 0$. The objective functions are the labor costs, the costs of the basic production assets and the costs of the manufactured products. The first and second objective functions have to be minimized; and so, we take them with minus sign to match the general statement of the decision-making problem where all criteria are maximized. Consequently, the criteria are defined by $f_1 = -p_1x_1$, $f_2 = -p_2x_2$, and $f_3 = p_zz$, where the quantities p_1, p_2, p_z specify the prices of the corresponding resources and manufactured products.

The set of feasible vectors Y represents the surface described by the equation $y_3 = \frac{ap_z}{p_1^{\alpha_1}p_2^{\alpha_2}}(-y_1)^{\alpha_1}(-y_2)^{\alpha_2}$ under the constraints $y_1, y_2 < 0$. The set of feasible vectors Y is shown by Fig. 8.4. As easily seen, this surface is concave and hence $P(Y) = Y$.

8.6.2 Pareto Set Reduction

Consider the two-dimensional sections of the set Y by the planes that are parallel to the coordinate planes (y_1, y_2), (y_2, y_3), (y_1, y_3). In the first case, we obtain the isoquant defined by the set

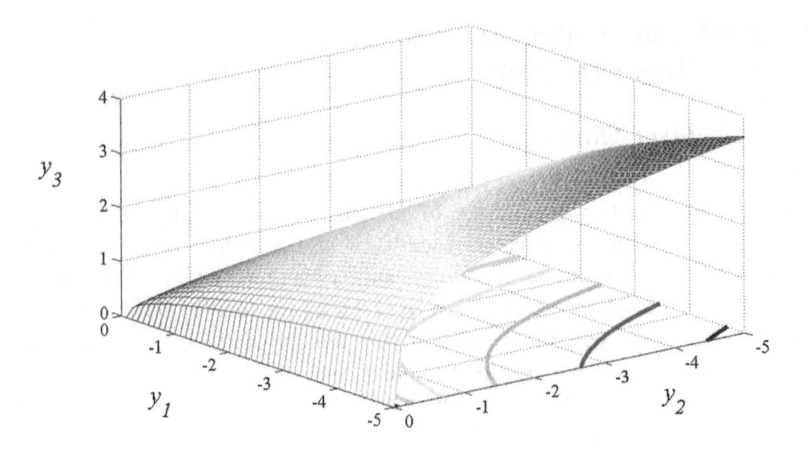

Fig. 8.4 Set Y

$$Y_{c_3}^{12} = \{(y_1, y_2) | (y_1, y_2, c_3) \in Y\}, (y_1, y_2).$$

This is a hyperbola for $c_3 > 0$ (see the set Y_4^{12} in Fig. 8.5). The DM reduces the costs of one resource only by increasing the costs of the other and conversely (note that the income remains the same). In other words, by choosing a point (y_1', y_2') instead of another point (y_1'', y_2''), where both points belong to the set $Y_{c_3}^{12}$ (and the corresponding points $(y_1', y_2', c_3), (y_1'', y_2'', c_3) \in P(Y)$), the DM admits a definite compromise between the labor costs and the costs of the basic production assets.

The sections of the set of feasible vectors Y by the planes $y_1 = c_1$ and $y_2 = c_2$, i.e.,

$$Y_{c_1}^{23} = \{(y_2, y_3) | (c_1, y_2, y_3) \in Y\}, c_1 < 0,$$

and

$$Y_{c_2}^{13} = \{(y_1, y_3) | (y_1, c_2, y_3) \in Y\}, c_2 < 0,$$

respectively, represent the power functions. Figure 8.6 shows these sets for $c_1 = -2$ and $c_2 = -3$.

Choosing a specific compromise alternative, the DM has to select the relationships between the income and each type of the costs.

Assume that the DM expresses its preferences in the following way.

1. Higher income f_3 is preferable to lower labor costs f_1. The DM is willing to compromise by increasing the labor costs (wages, etc.) by w_1^3 thousand RUB for gaining the additional income of w_3^3 thousand RUB, where $w_3^3 > w_1^3$.

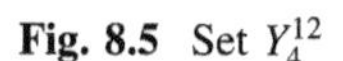
Fig. 8.5 Set Y_4^{12}

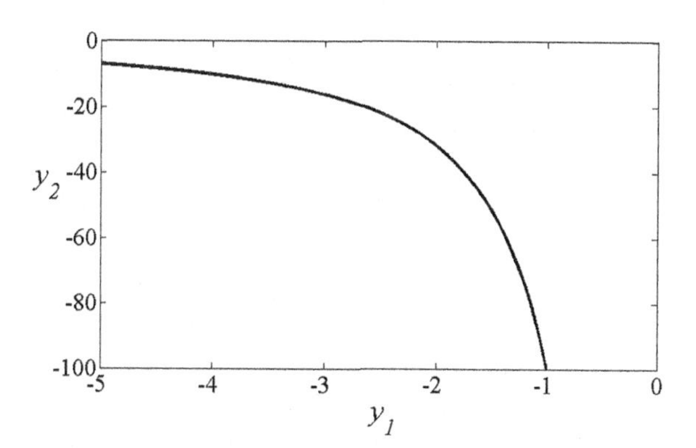

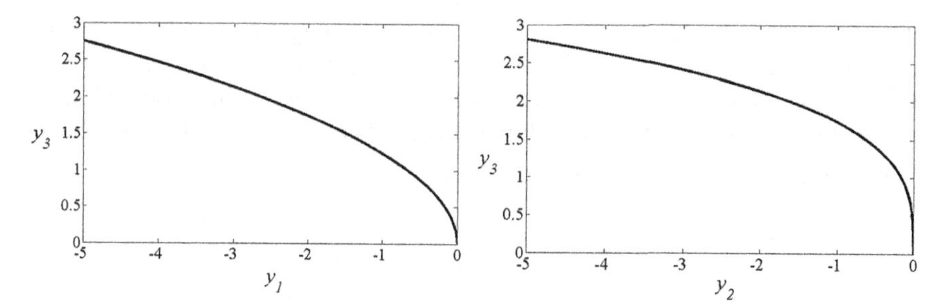

Fig. 8.6 Sets $Y_{c_1}^{32}$ and $Y_{c_2}^{13}$

2. Lower labor costs f_1 are preferable to lower costs of the basic production assets f_2. This DM uses high-quality (expensive) production equipment rather than high-quality labor. For reducing the labor costs by w_1^1 thousand RUB, the DM is willing to increase the costs of the basic production assets by w_2^1 thousand RUB, where $w_1^1 > w_2^1$.
3. Lower costs of the basic productions assets f_2 are preferable to higher income f_3. However, the purchased material, equipment, etc. (the basic production assets) must be reasonably priced. Therefore, the DM is willing to decrease the income by w_3^2 thousand RUB for reducing the costs of the basic production assets by w_2^2 thousand RUB, where $w_2^2 > w_3^2$.

Consequently, there exists the closed information that the labor costs f_1 have higher importance than the costs of the basic production assets f_2, the costs of the basic production assets f_2 have higher priority than the income f_3, and the income f_3 has higher importance than the labor costs f_1, with the corresponding collections of positive parameters. If we take into account the inequalities from items 1)-3), then the above information in the form of three quanta is consistent. Really,

$$|W| = \begin{vmatrix} w_1^1 & 0 & -w_1^3 \\ -w_2^1 & w_2^2 & 0 \\ 0 & -w_3^2 & w_3^3 \end{vmatrix} = w_1^1 w_2^2 w_3^3 - w_2^1 w_3^2 w_1^3 > 0.$$

Let us apply Corollary 5.2 to reduce the Pareto set. The new vector criterion g has the components

$$\begin{aligned} g_1 &= w_2^2 w_3^3 f_1 + w_1^3 w_3^2 f_2 + w_1^3 w_2^2 f_3 = -p_1 w_2^2 w_3^3 x_1 - p_2 w_1^3 w_3^2 x_2 + a p_z w_1^3 w_2^2 x_1^{\alpha_1} x_2^{\alpha_2}, \\ g_2 &= w_2^1 w_3^3 f_1 + w_1^1 w_3^3 f_2 + w_1^3 w_2^1 f_3 = -p_1 w_2^1 w_3^3 x_1 - p_2 w_1^1 w_3^3 x_2 + a p_z w_1^3 w_2^1 x_1^{\alpha_1} x_2^{\alpha_2}, \\ g_3 &= w_2^1 w_3^2 f_1 + w_1^1 w_3^2 f_2 + w_1^1 w_2^2 f_3 = -p_1 w_2^1 w_3^2 x_1 - p_2 w_1^1 w_3^2 x_2 + a p_z w_1^1 w_2^2 x_1^{\alpha_1} x_2^{\alpha_2}. \end{aligned}$$

The set of feasible alternatives X is convex. Under the accepted hypotheses and $\alpha_1 + \alpha_2 < 1$, the power function $q(x_1, x_2) = a p_z w_1^3 w_2^2 x_1^{\alpha_1} x_2^{\alpha_2}$ has concavity on X, which means that the function $g_1(x)$ is also concave for any $(x_1, x_2) \in X$ (as the sum of a power function and a linear function). By analogy we establish that the functions $g_2(x)$ and $g_3(x)$ are concave. In the final analysis, we obtain the problem with the convex set X and the concave vector criterion g. In this case, the set of proper efficient points (which is slightly narrower than the Pareto set $P_g(X)$) can be found using the linear combination of criteria

$$\begin{aligned} \varphi(x) = &-p_1 \left(\lambda_1 w_2^2 w_3^3 + \lambda_2 w_2^1 w_3^3 + \lambda_3 w_2^1 w_3^2\right) x_1 - p_2 \left(\lambda_1 w_1^3 w_3^2 + \lambda_2 w_1^1 w_3^3 + \lambda_3 w_1^1 w_3^2\right) x_2 \\ &+ a p_z \left(\lambda_1 w_1^3 w_2^2 + \lambda_2 w_1^3 w_2^1 + \lambda_3 w_1^1 w_2^2\right) x_1^{\alpha_1} x_2^{\alpha_2}. \end{aligned}$$

with the positive coefficients $\lambda_1, \lambda_2, \lambda_3$ such that $\lambda_1 + \lambda_2 + \lambda_3 = 1$. Denote

$$\begin{aligned} A_1 &= p_1 \left(\lambda_1 w_2^2 w_3^3 + \lambda_2 w_2^1 w_3^3 + \lambda_3 w_2^1 w_3^2\right), \\ A_2 &= p_2 \left(\lambda_1 w_1^3 w_3^2 + \lambda_2 w_1^1 w_3^3 + \lambda_3 w_1^1 w_3^2\right), \\ A_3 &= a p_z \left(\lambda_1 w_1^3 w_2^2 + \lambda_2 w_1^3 w_2^1 + \lambda_3 w_1^1 w_2^2\right). \end{aligned}$$

Consider the first partial derivatives of the linear combination

$$\varphi'_{x_1} = -A_1 + \alpha_1 A_3 x_1^{\alpha_1 - 1} x_2^{\alpha_2}, \ \varphi'_{x_2} = -A_2 + a_2 A_3 x_1^{\alpha_1} x_2^{\alpha_2 - 1}.$$

The solution of the system of equations $\varphi'_{x_1}(x^0) = 0$, $\varphi'_{x_2}(x^0) = 0$ yields the point $x^0 = (x_1^0, x_2^0)$ defined by

$$\begin{aligned} x_1^0 &= \left(\frac{\alpha_1 \alpha_3}{A_1}\right)^{\frac{1-\alpha_2}{1-(\alpha_1+\alpha_2)}} \left(\frac{\alpha_2 \alpha_3}{A_2}\right)^{\frac{\alpha_2}{1-(\alpha_1+\alpha_2)}}, \\ x_2^0 &= \left(\frac{\alpha_1 \alpha_3}{A_1}\right)^{\frac{\alpha_1}{1-(\alpha_1+\alpha_2)}} \left(\frac{\alpha_2 \alpha_3}{A_2}\right)^{\frac{1-\alpha_2}{1-(\alpha_1+\alpha_2)}}, \end{aligned}$$

where $x_1^0, x_2^0 > 0$.

Now, calculate the second partial derivatives of the linear combination at the point $x = x^0$:

$$\varphi''_{x_1^2}(x^0) = \alpha_1 A_3(\alpha_1 - 1)(x_1^0)^{\alpha_1 - 2}(x_2^0)^{\alpha_2},$$
$$\varphi''_{x_2^2}(x^0) = \alpha_2 A_3(\alpha_2 - 1)(x_1^0)^{\alpha_1}(x_2^0)^{\alpha_2 - 2},$$
$$\varphi''_{x_1 x_2}(x^0) = \alpha_1 \alpha_2 A_3 (x_1^0)^{\alpha_1 - 1}(x_2^0)^{\alpha_2 - 1}.$$

The function $\varphi(x)$ reaches its maximum at the point $x = x^0$ under the sufficient conditions

$$\varphi''_{x_1^2}(x^0) < 0, \quad \varphi''_{x_1^2}(x^0) \cdot \varphi_{x_2^2}(x^0) - (\varphi''_{x_1 x_2}(x^0))^2 > 0.$$

As easily seen, the first inequality holds owing to $\alpha_2, A_3, x_1^0, x_2^0 > 0$, and $0 < \alpha_1 < 1$. For the second inequality, we have

$$\varphi''_{x_1^2}(x^0) \cdot \varphi''_{x_2^2}(x^0) - (\varphi''_{x_1 x_2}(x^0))^2 = A_3^2 \alpha_1 \alpha_2 (1 - (\alpha_1 + \alpha_2))(x_1^0)^{2(\alpha_1 - 1)}(x_2^0)^{2(\alpha_2 - 1)} > 0,$$

since $\alpha_1, a_2, x_1^0, x_2^0 > 0$, and $\alpha_1 + \alpha_2 < 1$. Thus, the point $x = x^0$ is the maximum point of the linear combination $\varphi(x)$ and hence it is proper efficient. As x^0 depends on the positive parameters λ_1, λ_2, and λ_3, we have actually constructed a whole family of such points. The final choice must be performed exactly within this set.

8.7 Weakening of Basic Axiomatics

As a rule, in practice one does not know whether Axioms 1–4 hold or not. Therefore, a challenging problem is to weaken these axiomatics in order to enlarge the associated class of multicriteria choice problems that can be solved using the axiomatic approach with Pareto set reduction based on information quanta.

Below we discuss two possible directions of such weakening.

8.7.1 Weakening of Compatibility Axiom

It has been established in Chap. 2 (see Lemma 2.2 and Theorem 2.1) that a preference relation satisfying Axioms 2 and 4 is a cone relation with an acute convex cone not containing the origin. If we also accept Axiom 3, then this cone contains the nonnegative orthant. Consequently, it is possible to suggest the following (generalized) statement of the compatibility axiom.

Axiom 3′ *In the criterion space R^m there is a given cone of "desired directions" C that is acute, convex and does not contain the origin. Moreover, for each pair of vectors $y', y'' \in R^m$ the inclusion $y' - y'' \in C$ implies the relationship $y' \succ y''$.*

In Axiom 3 the role of the cone C is played by the nonnegative orthant of space R^m. In the general case, according to Axiom 3′ the cone C may be wider or narrower than the nonnegative orthant. The last axiom can be treated as some generalization of the Pareto axiom.

By accepting Axiom 3′, we have to modify the definition of an information quantum in the following way.

Definition 8.1 Consider a certain pair of vectors $y', y'' \in R^m$ that satisfy neither the relationship $y' - y'' \in C$ nor the relationship $y'' - y' \in C$. Under Axiom 3′, we say that *there is a given information quantum about the DM's preference relation* if one of the relationships $y' \succ y''$ or $y'' \succ y'$ holds.

Then the following result is true.

Theorem 8.10 *Under Axioms* 1, 2, 3′ *and* 4, *assume that there exists an information quantum (in the sense of Definition* 8.1) *stating that $y' \succ y''$. Then for any set of selectable vectors C(Y) we have the inclusion*

$$C(Y) \subset \mathrm{Ndom}_C Y \tag{8.18}$$

where

$$\mathrm{Ndom}_C Y = \{y^* \in X \mid there\ exists\ no\ y \in Y\ such\ that\ y - y^* \in conv\{(y' - y'') \cup C\}\},$$

and conv{A} denotes the convex hull of the set A. Moreover, if the cone C contains the nonnegative orthant, then we also have the inclusion $\mathrm{Ndom}_C Y \subset P(Y)$.

□ As mentioned in the beginning of this subsection, a preference relation satisfying Axioms 2 and 4 is a cone relation with an acute convex cone K not containing the origin. Owing to Axiom 3′ we have the inclusion $C \subset K$. Besides, the presence of the information quantum gives $K \supset conv\{(y' - y'') \cup C\}$. Hence, $\mathrm{Ndom}\, Y \subset \mathrm{Ndom}_C\, Y$. This result together with the inclusion $C(Y) \subset \mathrm{Ndom}\, Y$ (which follows from Axiom 1) brings to the required inclusion (8.18). ■

Theorem 8.10 can be extended to the case of a consistent collection of information quanta defined by the pairs of vectors $u^i, v^i \in R^m$, $u^i - v^i \notin \{C \cup (-C)\}, i = 1, 2, \ldots, k$. Prior to this extension, let us formulate the corresponding definition of a consistent collection of information quanta.

Definition 8.2 Under Axiom 3′, we say that a collection of the pairs of vectors $u^i, v^i \in R^m$, $u^i - v^i \notin \{C \cup (-C)\}, i = 1, 2, \ldots, k$, is *consistent* if there exists an irreflexive binary relation $\succ$ satisfying Axioms 2 and 4 such that the relationships $u^i \succ v^i$, $i = 1, \ldots, k$, hold.

Theorem 8.11 *Accept Axiom 3′. A collection of the pairs of vectors* $u^i, v^i \in R^m$, $u^i - v^i \notin \{C \cup (-C)\}, i = 1, 2, \ldots, k,$ *is consistent if the convex hull* $conv\left\{\left[\bigcup_{i=1,\ldots,k}(u^i - v^i)\right] \cup C\right\}$ *forms an acute cone.*

□ Indeed, the cone relation with the acute convex cone $conv\left\{\left[\bigcup_{i=1,\ldots,k}(u^i - v^i)\right] \cup C\right\}$ satisfies Axioms 2 and 4 by Theorem 2.1. ■

For the corresponding generalization of Theorem 8.10 (that holds for an arbitrary consistent collection of the pairs of vectors $u^i, v^i \in R^m$, $u^i - v^i \notin \{C \cup (-C)\}, i = 1, 2, \ldots, k$), in the definition of the set $\text{Ndom}_C\, Y$ the convex hull $conv\{(y' - y'') \cup C\}$ is replaced by the convex hull $conv\left\{\left[\bigcup_{i=1,\ldots,k}(u^i - v^i)\right] \cup C\right\}$.

If the set of feasible vectors Y is finite, then the set $\text{Ndom}_C\, Y$ can be in principle constructed by the direct enumeration of all pairs of vectors from the set Y. However, complex computational problem may arise in the case of the infinite set Y. We will not discuss the details here.

In the elementary situation (the cone C is polyhedral), the set $\text{Ndom}_C\, Y$ can be constructed by a similar algorithm as the one used for dual cone design (see Sect. 5.3).

8.7.2 Weakening of Invariance Axiom

As we have emphasized above, a preference relation satisfying Axioms 2 and 4 is a cone relation with an acute convex cone not containing the origin. The idea of the next generalization of the axiomatic approach is to reject the linearity (more specifically, homogeneity) of the preference relation and to postulate the existence of some "good" cone within the set of all vectors dominating an arbitrary vector of the criterion space. In particular, suppose that instead of Axioms 2-4 the preference relation $\succ$ obeys the following requirement.

Axiom 2′′ *The preference relation* $\succ$ *is additive and, in addition, for each* $y \in R^m$ *the set* $Y_y = \{z \in R^m | y \succ z\}$ *is convex and contains some fixed acute convex cone* C *without the origin.*

Definition 8.3 Consider a certain pair of vectors $y', y'' \in R^m$ that satisfy neither the relationship $y' - y'' \in C$ nor the relationship $y'' - y' \in C$. Under Axiom 2′′, we say that *there is a given information quantum* if one of the relationships $y' \succ y''$ or $y'' \succ y'$ holds.

The elicitation problem of such information quanta turns out much more difficult than the corresponding problem based on Definition 8.1, since the cone C is initially unknown.

The following result takes place.

Theorem 8.12 *Under Axioms* 1 *and* $2''$, *assume that there exists an information quantum (in the sense of Definition* 8.3) *stating that* $y' \succ y''$. *Then for any set of selectable vectors* $C(Y)$ *we have inclusion* (8.18) *with the set described by Theorem* 8.11 *in the right-hand side. Moreover, if the cone* C *contains the nonnegative orthant, then we also have the inclusion* $\mathrm{Ndom}_C\, Y \subset P(Y)$.

This result follows from Axiom 1 and the fact that the convex hull $conv\{(y' - y'') \cup C\}$ is contained in the set $Y_y = \{z \in R^m | y \succ z\}$ for each $y \in R^m$. It can be extended to the case of an arbitrary finite consistent collection of information quanta, just like in the previous subsection.

Concluding Remarks

It is considered that French mathematician and political scientist J.-C. de Borda first attempted to solve the multicriteria problem. In particular, he suggested the ranked preferential voting system with the linear combination of criteria, see [1]. In the middle of the 19th century Irish economist F. Edgeworth [8] introduced the so-called "Edgeworth box," which actually involved the notion of a locally Pareto optimal alternative in terms of two criteria long before V. Pareto. The general notion of Pareto optimality appeared at the junction of the 19th and 20th centuries, but its intensive usage started in the 1940s–1950s. The research of that period was mostly dedicated to different generalizations of the well-known results on optimization theory, i.e., the development of necessary and sufficient optimality conditions and also existence conditions for certain optimality concepts, as well as to the duality issues in multicriteria programming (the problems with constraints defined by the solution set of system of equalities and/or inequalities). Nowadays this direction of investigations continues its evolvement, parallel to the corresponding branches of single-criterion optimization theory. In this context, also mention the research works suggesting different algorithms (including approximate ones) for Pareto set construction. For example, for the linear problems was developed the multicriteria analog of the simplex method, which yields all facets of the Pareto set.

On the other hand, following the vital demands of economics and engineering and the associated multicriteria optimization problems, many authors started suggesting different "best" solutions of the multicriteria problems using certain heuristic considerations. The pioneering results in this field belong to de Borda, see above. The scientific literature of the 1970s–1980s provides numerous examples illustrating how the linear combination of criteria (and other scalarization methods) can be used to solve various economic and engineering problems. In the 1980s it became finally clear that the "best" alternative choice cannot be justified without involving additional information (not including the collection of criteria and the set of feasible alternatives). That period was remarkable for the development and usage of the so-called "decision rules" that allow to extract the "best" solutions of the multicriteria problems in a certain sense. In the USSR (by then, with a considerable community of researchers focused on multicriteria optimization), different authors

© Springer International Publishing AG 2018

V.D. Noghin, *Reduction of the Pareto Set*, Studies in Systems, Decision and Control 126, https://doi.org/10.1007/978-3-319-67873-3

introduced decision rules by designing some "resulting" binary relations. Note that a similar trend showed up since the 1950s in the western countries after the appearance of Arrow's impossibility theorem (a Nobel Prize winner in economics). Subsequently, this trend yielded the general theory of alternative choice. An endeavor to translate the result of this theorem into the multicriteria language gives the following: generally, the multicriteria problem is not reducible to the single-criterion problem, since these problems are qualitatively different. After the impossibility theorem, hundreds of papers continued further analysis of the theoretical aspects and constructive ways to "aggregate" some general relation from a finite collection of partial binary relations. In terms of multicriteria optimization this means the reduction of the multicriteria problem to the single-criterion one (i.e., the scalarization of the multicriteria problem).

As mentioned, in the 1980s it became clear that the "best" alternative choice cannot be justified without involving additional information. For instance, such information may specify certain parameters (e.g., the weight coefficients of the linear combination of criteria) that participate in the corresponding scalarization approach. A series of authors proposed "the best alternative" based on some analogies or general considerations. As an example, refer to the center of gravity for the set of Pareto optimal vectors (by analogy with the Shapley value from game theory) or the Pareto optimal vector having the shortest distance to a certain ideal unattainable vector (like in goal programming).

According to the gradually maturing idea, the researchers started believing that the final choice is performed by an individual interested in the solution of the multicriteria problem (called the decision-maker). Each human has the right to consider its own "best" alternatives. Therefore, all attempts to suggest a universal rule or notion of the "best" alternative are doomed from the start. Embedding the binary preference relation in the multicriteria problem gave an opportunity to take into account the specifics of certain DM. However, the difficulty is that a human dealing with the choice problem often has a hazy idea of its preferences. In any case, the DM is unable to describe completely its preference relation. And the path of further development laid towards the consideration of some "fragmentary" information about the DM's preference relation, with minimum assumptions imposed on it. Such information was represented by the pairs of incomparable vectors in terms of the Pareto relation: in each pair, the DM surely prefers one vector to the other. And such information was later called the information quanta about the DM's preference relation.

The preference relation was explicitly incorporated into the multicriteria problem statement in the author's report presented in 1982 ([29]). The cited report also included the requirements in the form of axioms imposed on this relation (irreflexivity, the Pareto axiom, transitivity and invariance). The role of additional information about the DM's preference relation was played by a finite collection of the pairs of incomparable vectors (in terms of the Pareto relation) where one vector is preferable to the other. The axiomatic approach originated in this report. Later on, the axiomatic approach was developed in detail first in 1986 for the bicriteria problems [30] and then in 1991 for the multicriteria problems with an arbitrary finite number of criteria [31]. The multicriteria problem supplemented by the DM's binary preference relation was subsequently called the multicriteria choice problem in order to underline its connection to general choice theory (especially, to the paired dominant choice, i.e., the choice based on a certain binary relation). Next, the theorem on taking into account a general information quantum was proved in [33], and the Edgeworth-Pareto principle was logically justified in [37].

Year 2003 saw the monograph by the author on the quantitative approach to decision-making in multicriteria environment, which systematized the results obtained by then. The second edition of the monograph was published in 2005, see [38]. As a matter of fact, a series of new interesting results was published since that time. Presently, the axiomatic approach to reduce the Pareto set can be considered well-developed and, as the author believes, this book gives a rather complete description of the theory.

The interconnection between the original axiomatic approach and other methods and approaches was discussed in [39].

In conclusion, note that in some papers (see, for example, [16]) the authors avoid the axiomatic characterization of the preference relation and operate the terminology of cone-based approach with allowable tradeoffs among criteria. In fact, they deal with information quanta in the special case where all parameters w_i are 1.

References

1. Aizerman, M., & Aleskerov, F. (1995). *Theory of Choice*. North-Holland: Elsevier.
2. Arrow, K. J., Barankin, E. W., & Blackwell, D. (1953). Admissible points of convex sets. In H. W. Kuhn & A. W. Tucker (Eds.), *Contributions to the theory of games* (pp. 87–91). Princeton: Princeton University Press.
3. Baskov, O. V. (2015). An algorithm for Pareto set reduction using fuzzy information on decision maker's preference relation. *Scientific and Technical Information Processing, 42*(5), 382–387.
4. Baskov, O. V. (2014). Kriteriy neprotivorechivosti «kvantov» informatsii o nechetkom otnoshenii predpochteniya litsa, prinimayuschego resheniya (A criterion of consistency of information quanta on the preference relation of the decision maker). *Vestn. St. Petersburg Univ. Ser. 10: Applied Math. Comp. Sci. Contr. Processes, 2*, 13–19 (in Russian).
5. Borwein, J. (1977). Proper efficient points for maximization with respect to cones. *SIAM Journal on Control and Optimization, 15*(1), 57–63.
6. Charns, A., & Cooper, W. W. (1961). *Management models and industrial applications of linear programming (Appendix B)*. New York: Wiley.
7. Chernikov, S. N. (1968). *Lineinye neravenstva* (Linear inequalities) (p. 352). Moscow: Nauka (in Russian).
8. Edgeworth, F. Y. (1908). Appreciations of mathematical theories. *Economic Journal, 18*(72), 541–556.
9. Ehrgott, M. (2005). *Multicriteria optimization*. Berlin-Heidelberg: Springer.
10. Figueira J., Greco S., & Ehrgott M., (Eds.). (2005). *Multiple criteria decision analysis: state of the art surveys*. Springer's International Series in Operations Research and Management Science. New York, NY: Springer Science + Business Media.
11. Fishburn, P. C. (1970). *Utility theory for decision making*, Publications in Operations Research (Vol. 18). New York: Wiley.
12. Fishburn, P. C. (1964). *Decision and value theory*, Publications in Operations Research (Vol. 10). New York: Wiley.
13. Gearhart, W. B. (1979). Compromise solutions and estimation of noninferior set. *Journal of Optimization Theory and Applications, 28*(1), 29–47.
14. Geoffrion, A. M. (1968). Proper efficiency and the theory of vector maximization. *Journal of Mathematical Analysis and Applications, 22*(3), 618–630.
15. Germeier, Yu. B. (1967). *Metodologicheskie i matematicheskie osnovyi issledovaniya operatsiy i teorii igr* (*Methodological and mathematical foundation of operation research and game theory*). Moscow: Computer Center of Moscow State University (in Russian).
16. Hunt, B. J., Wiecek, M. M., & Hughes, C. S. (2010). Relative importance of criteria in multiobjective programming: A cone-based approach. *European Journal of Operational Research, 207*, 936–945.
17. Jhan, J. (1987). Parametric approximation problems arising in vector optimization. *Journal of Optimization Theory and Applications, 54*(3), 503–516.

© Springer International Publishing AG 2018

V.D. Noghin, *Reduction of the Pareto Set*, Studies in Systems,
Decision and Control 126, https://doi.org/10.1007/978-3-319-67873-3

18. Jahn, J. (2004). *Vector optimization. Theory, applications, and extensions* (p. 463). Berlin: Springer.
19. Keeney, R. L., & Raiffa, H. (1976). *Decisions with multiple objectives: preferences and value tradeoffs*. New York: Wiley.
20. Kaufmann, A. (1975). *Introduction to the theory of fuzzy subsets* (p. 416). New York: Academic Press.
21. Klir, G. J., St Clair, U., Yuan, B. (1997). *Fuzzy Set Theory: Foundations and Applications*. Upper Saddle River, NJ: Prentice Hall.
22. Kowalczyk, C., & Skeath, S. E. (1994). Pareto ranking optimal tariffs under foreign monopoly. *Economics Letters, 45,* 355–359.
23. Leichtweiss, K. (1980). *Konvexe Mengen*. Berlin: Springer.
24. Lotov, A. V., Bushenkov, V. A., & Kamenev, G. K. (2004). *Interactive decision maps. Approximation and visualization of Pareto frontier* (p. 336). Boston: Kluwer Academic Publishers.
25. Mangasarian, O. (1969). *Nonlinear Programming*. New York: McGraw Hill.
26. Miettinen, K. (1999) *Nonlinear multiobjective optimization*, Vol. 12 of International Series in Operations Research and Management Science. Kluwer Academic Publishers, Dordrecht.
27. Mendeleev, D. I. (1950). Tolkovyiy tarif ili issledovanie o razvitii promyishlennosti Rosii v svyazi s ee obschim tamozhennyim tarifom 1891 goda (Reasonable tariff, or the analysis of industrial development in Russia in the context of its general customs tariff established in 1891), in *Complete Works of D. I. Mendeleev*, (Vol. 19, pp. 230–937) (in Russian).
28. Miller, G. A. (1956). The magical number seven, plus or minus two: Some limits on our capacity for processing information. *Psychological Review, 63*(2), 81–97.
29. Noghin, V. D. (1983). Otsenki dlya mnozhestva optimalnyih resheniy v usloviyah otnosheniya predpochteniya, invariantnogo otnositelno polozhitelnogo lineynogo preobrazovaniya (Estimates for optimal solution set under preference relation invariant with respect to positive linear transformation). In *Proceedings of all-union conference "Decision-making under multiple criteria and uncertainty*," (p. 37). Moscow–Batumi (in Russian).
30. Noghin, V. D. et al. (1986). *Osnovy teorii optimizatsii* (*Fundamentals of optimization theory*) (p. 384). Moscow: Vysshaya Shkola, (in Russian).
31. Noghin, V. D. (1991). Estimation of the set of nondominated solutions. *Numerical Functional Analysis and Applications, 12*(5–6), 507–515.
32. Noghin, V. D. (1994). Upper estimate for a fuzzy set of nondominated solutions. *Fuzzy Sets and Systems, 67,* 303–315.
33. Noghin, V. D. (1997). Relative importance of criteria: A quantitative approach. *Journal of Multi-Criteria Decision Analysis, 6,* 355–363.
34. Noghin, V. D. (2000). Ispolzovanie kolichestvennoy informatsii ob otnositelnoy vazhnosti kriteriev v prinyatii resheniy (Using quantitative information about the relative importance of criteria in decision-making). *Nauchno-tekhnicheskie Vedomosti Sankt-Peterburg. Gos. Tekh. Univ.* (Vol. 1, pp. 89–94) (in Russian).
35. Noghin, V. D. (2000). Teoremyi o polnote v teorii otnositelnoy vazhnosti kriteriev (Completeness theorems in theory of the relative importance of criteria). *Vestn. St. Petersburg Univ. Ser. 1: Mathematics. Mechanics. Astronomy, 40*(25), 13–18. (in Russian).
36. Noghin, V. D. (2001). What is the relative importance of criteria and how to use it in MCDM. In Köksalan, M. & Zionts, S. (Eds.). *Multiple Criteria Decision Making in the New Millenium, Proceedings of the XV International Conference on MCDM* (pp. 59–68). Ankara, Turkey, July 2000; Springer.
37. Noghin, V. D. (2002). Logicheskoe obosnovanie printsipa Edzhvorta-Pareto (A logical justification of the Edgeworth-Pareto principle). *Computational Mathematics and Mathematical Physics, 42*(7), 951–957 (in Russian).
38. Noghin, V. D. (2005). *Prinyatie reshenii v mnogokriterial'noi srede: kolichestvennyi podkhod* (Decision-making in multicriteria environment: Quantitative approach). Moscow: Fizmatlit, (in Russian).

39. Noghin, V. D. (2008). Problema suzheniya mnozhestva Pareto: podhodyi k resheniyu (Problem of Pareto set reduction: solution approaches). *Artificial Intelligence and Decision Making, 1,* 98–112 (in Russian).
40. Noghin, V. D. (2010). Restriction of a Pareto set based on information about decision-maker's preferences of the point-multiple type. *Scientific and Technical Information Processing, 37*(5), 292–300.
41. Noghin, V. D. (2011). Reducing the Pareto set based on set-point information. *Scientific and Technical Information Processing, 38*(6), 435–439.
42. Noghin, V. D. (2012). Pareto set reduction based on fuzzy information. *International Journal "Information Technologies & Knowledge", 6*(2), 157–168.
43. Noghin, V. D. (2013). An axiomatic approach to Pareto set reduction: computational aspects. *International Journal "Information Theories and Applications", 20*(4), 352–359.
44. Noghin, V. D. (2014). Reducing the Pareto set algorithm based on an arbitrary finite set of information "quanta". *Scientific and Technical Information Processing, 41*(5), 309–313.
45. Noghin, V. D. (2015). Linear scalarization of criteria in multi-criterion optimization. *Scientific and Technical Information Processing, 42*(6), 463–469.
46. Noghin, V. D. (2015). Generalized Edgeworth-Pareto principle. *Computational Mathematics and Mathematical Physics, 55*(12), 1975–1980.
47. Noghin, V. D. (2017). Pareto set reduction based on an axiomatic approach with application of some metrics. *Computational Mathematics and Mathematical Physics, 57*(4), 645–652.
48. Noghin, V. D., & Baskov, O. V. (2011). Pareto set reduction based on an arbitrary finite collection of numerical information on the preference relation. *Doklady Mathematics, 83*(3), 418–420.
49. Noghin, V. D., & Tolstykh, I. V. (2000). Ispolzovanie kolichestvennoy informatsii ob otnositelnoy vazhnosti kriteriev v prinyatii resheniy (Using quantitative information on the relative importance of criteria for decision making). *Computational Mathematics and Mathematical Physics, 40*(11), 1529–1536. (in Russian).
50. Noghin, V. D., & Prasolov, A. V. (2013). Mnogokriterialnaya otsenka optimalnoy velichinyi tamozhennoy poshlinyi (Multicriteria evaluation of the optimal amount of the import duty). *Trudy Inst. Sistemn. Analiza, 2,* 34–44. (in Russian).
51. Orlovsky, S. A. (1980). On formalization of a general fuzzy mathematical programming. *Fuzzy sets and systems, 3*(3), 311–321.
52. Pareto V. (1896). *Manual d''economie politique*. F. Rouge, Lausanne (in French).
53. Petrovskii, A. B. (2009). *Teoriya prinyatiya reshenii* (Theory of decision-making) (p. 400). Moscow: Akademiya (in Russian).
54. Plous, S. (1993). *The psychology of judgment and decision making* (p. 302). McGraw-Hill.
55. Podinovskii, V. V. & Noghin, V. D. (2007). *Pareto-optimal'nye resheniya mnogokriterial'nykh zadach* (*Pareto optimal solutions of multicriteria problems*, 2nd ed.) (p. 256). Moscow: Fizmatlit, (in Russian).
56. Prasolov, A. V. (1999). Ob odnom vozmozhnom podhode k analizu protektsionistskoy politiki (On one of possible approaches to analysis of protectionist policy). *Economics and Mathematical Methods, 35*(2), 153–156 (in Russian).
57. Rockafellar, R. T. (1970). *Convex Analysis*. Princeton: Princeton University Press.
58. Saaty, T. L. (1990). *Multicriteria decision making. The analytic hierarchy process* (287 p). Pittsburgh: RWS Publications,
59. Stadler, W. (1979). A survey of multicriteria optimization or the vectormaximum problem, Part I: 1776–1960. *Journal of Optimization Theory and Applications, 29,* 1–52.
60. Steuer, R. (1985). *Multiple criteria optimization: Theory, computation and application*. New York: Wiley.
61. Tamura, K. (1976). A method for constructing the polar cone of a polyhedral cone, with applications to linear multicriteria decision problems. *Journal of Optimization Theory and Applications, 19*(4), 547–564.
62. Schrijver, A. (1998). *Theory of linear and integer programming* (484 p). Wiley.

63. *The Tax Code of the Russian Federation*, Section VIII: Federal taxes; chapter 21: Value-added tax. http://base.garant.ru/10900200/28/#20021 2002.
64. *Trends in multiple criteria decision analysis* (Greco S., Ehrgott M., Figueira J., eds.). Springer, 2010, New York, 412 p.
65. Yu, P. L. (1985). *Multiple-criteria decision making: concepts, techniques, and extensions* 388 pp. New York, London: Plenum Press.
66. Zadeh, L. A. Fuzzy sets. *Information and Control*, *8*(3), 338–353.
67. Zakharov, A. O. (2009). Suzhenie mnozhestva Pareto na osnove zamknutoy informatsii ob otnoshenii predpochteniya (Reduction of the Pareto set based on closed information of the DM's preference relation). *Vestn. St. Petersburg Univ. Ser. 10: Applied Math. Comp. Sci. Contr. Processes*, *4*, 69–83 (in Russian).
68. Zakharov, A. O. (2011). Suzhenie mnozhestva Pareto na osnove zamknutoy informatsii o nechetkom otnoshenii predpochteniya litsa, prinimayuschego reshenie (Pareto set reduction based on interdependent information of closed type). *Artificial Intelligence and Decision Making 1*, 67–81 (in Russian).
69. Zakharov, A. O. (2012). Suzhenie mnozhestva Pareto na osnove zamknutoy informatsii o nechetkom otnoshenii predpochteniya litsa, prinimayuschego reshenie (Reduction of the Pareto set based on closed information about fuzzy DM's preference relation). *Vestn. St. Petersburg Univ. Ser. 10: Applied Math. Comp. Sci. Contr. Processes*, *3*, 33–47 (in Russian).
70. Zakharov, A. O. (2012). Pareto set reduction using compound information of a closed type. *Scientific and Technical Information Processing*, *39*(5):293–30.

Index

© Springer International Publishing AG 2018

V.D. Noghin, *Reduction of the Pareto Set*, Studies in Systems, Decision and Control 126, https://doi.org/10.1007/978-3-319-67873-3

Printed in the United States
By Bookmasters